# How Does the Weapon Industry Work

CRAFTED BY SKRIUWER

**Copyright © 2025 by Skriuwer.**

All rights reserved. No part of this book may be used or reproduced in any form whatsoever without written permission except in the case of brief quotations in critical articles or reviews.

At **Skriuwer**, we're more than just a team—we're a global community of people who love books. In Frisian, "Skriuwer" means "writer," and that's at the heart of what we do: creating and sharing books with readers worldwide. Wherever you are in the world, **Skriuwer** is here to inspire learning.

**Frisian** is one of the oldest languages in Europe, closely related to English and Dutch, and is spoken by about **500,000 people** in the province of **Friesland** (Fryslân), located in the northern Netherlands. It's the second official language of the Netherlands, but like many minority languages, Frisian faces the challenge of survival in a modern, globalized world.

We're using the money we earn to promote the Frisian language.

For more information, contact : **kontakt@skriuwer.com** (www.skriuwer.com)

# TABLE OF CONTENTS

## CHAPTER 1: UNDERSTANDING WHAT WEAPONS ARE

- *What a weapon is and why people use it*
- *How weapons are different from regular tools*
- *Simple examples of weapons around the world*

## CHAPTER 2: A BRIEF LOOK AT THE HISTORY OF WEAPONS

- *Early tools that turned into weapons*
- *How weapons changed over time*
- *Famous weapon inventions in history*

## CHAPTER 3: HOW MODERN WEAPONS ARE DESIGNED

- *Steps in creating new weapons*
- *Materials and technology used*
- *Why testing is important*

## CHAPTER 4: SMALL ARMS AND THEIR ROLE

- *What small arms are and who uses them*
- *Ways small arms help in defense*
- *Key differences among types of small arms*

## CHAPTER 5: LARGE-SCALE GUNS AND CANNONS

- *How big guns and cannons work*
- *Where they are used*
- *Reasons why they are still important*

## CHAPTER 6: TANKS AND ARMORED VEHICLES

- What makes tanks and armored vehicles special
- How they protect people inside
- Common features and uses

## CHAPTER 7: NAVAL SHIPS AND THEIR WEAPONS

- Types of ships used by navies
- Weapons found on board
- How these ships help in defense

## CHAPTER 8: AIRCRAFT AND THEIR WEAPON SYSTEMS

- Different types of military planes
- How weapons are added to aircraft
- Why speed and height matter

## CHAPTER 9: MISSILES AND ROCKETS

- How missiles and rockets are guided
- Reasons they are powerful
- Where they might be used

## CHAPTER 10: NUCLEAR WEAPONS AND THEIR EFFECTS

- Basic facts about nuclear weapons
- Why they are so harmful
- Past examples of nuclear use

## CHAPTER 11: WHO MAKES WEAPONS AND WHY

- *Companies and groups involved*
- *Reasons for creating and selling weapons*
- *People who work in these industries*

## CHAPTER 12: RESEARCH AND DEVELOPMENT IN THE WEAPON INDUSTRY

- *How new ideas are tested*
- *Places where research takes place*
- *Why improvements never stop*

## CHAPTER 13: MONEY MATTERS: FUNDING AND PROFITS

- *Where weapon funds come from*
- *How big the profits can be*
- *Effects on the economy*

## CHAPTER 14: HOW WEAPONS ARE SOLD AROUND THE WORLD

- *Types of deals between countries*
- *Why weapons are shipped far away*
- *Rules that try to control sales*

## CHAPTER 15: SUPPLY CHAINS AND DISTRIBUTION

- *Steps from factory to battlefield*
- *Transport methods used*
- *Problems that can slow down delivery*

## CHAPTER 16: GOVERNMENT RULES AND OVERSIGHT

- *How governments try to control weapons*
- *Why there are licenses and checks*
- *Ways rules differ by country*

## CHAPTER 17: SAFETY AND SECURITY CONCERNS

- *Why accidents happen and how to avoid them*
- *Protecting people from weapon risks*
- *How stolen weapons can be a threat*

## CHAPTER 18: EFFECTS ON CONFLICTS AND WARS

- *How weapons shape battles*
- *Why advanced tools can change outcomes*
- *What happens after wars end*

## CHAPTER 19: QUESTIONS ABOUT ETHICS AND MORALS

- *Thinking about right and wrong*
- *Why people argue about weapon use*
- *Ways to solve disagreements*

## CHAPTER 20: POSSIBLE FUTURES OF THE WEAPON INDUSTRY

- *How technology might change weapons*
- *What may happen if rules get stronger*
- *Why people still wonder about the future*

# CHAPTER 1

## UNDERSTANDING WHAT WEAPONS ARE

A weapon is an object or tool made to harm or protect. It might sound strange to call it both harmful and protective, but it depends on who is using it and why. When a person has a weapon, they can attack someone else, but they can also defend themselves or their group from danger. Because of this, weapons have a big effect on our world. Some are simple, like a sharpened stick or a stone, while others are very advanced, like modern firearms or even bigger systems. In this chapter, we will look at what weapons are, how they differ from normal tools, and why people feel a need to have them. We will also talk about examples of different kinds of weapons around the world.

## What Is a Weapon?

A weapon is not just any object that can harm someone. For example, a rock can cause harm, but if a rock is just lying on the ground, it is not called a weapon. Once a person picks up that rock and plans to use it to hurt or protect, it becomes a weapon. This is because the design and the use both matter.

Tools like hammers and kitchen knives can also be used to hurt someone, but we do not usually call them weapons. Why? Because their main job is not to harm people; it is to help with building or cooking. A hammer helps push nails into wood, and a kitchen knife helps cut vegetables or meat. But if a hammer or a kitchen knife is used to attack someone, it can be seen as a weapon in that situation.

Weapons are mostly made for one main purpose: to have an advantage over a threat, whether it is a wild animal, an enemy, or any danger. This makes them a special category among objects we use in everyday life.

## Why People Use Weapons

People may use weapons for many reasons. One reason is to protect themselves. Imagine you are in a place where there are dangerous animals. Having a strong stick might help you scare them away. Another reason is to hunt. Long ago, people made bows and arrows to catch animals for food. That made it easier to hunt than chasing animals by hand.

Some people also use weapons for power or control. In history, groups that had better weapons often won battles against groups with weaker weapons. This was because better weapons allowed them to strike harder or from a safer distance. Over time, having stronger weapons became a key part of many armies around the world.

There are also people who own weapons for sporting events, like target shooting. While these weapons can still cause harm, their main purpose in sports is accuracy and skill, not to hurt others.

## Differences Between Weapons and Regular Tools

1. **Purpose**:
   Weapons are mainly built to harm or protect. Regular tools usually help us with work, like cutting wood, cooking, or building houses.
2. **Design**:
   A lot of weapons have special parts meant only for fighting. For example, a sword has a sharp blade along its entire

length, and a gun has a barrel that fires bullets. Though some tools share shapes or parts with weapons, they are usually designed to do tasks in a safe way.

3. **Legal Rules**:
Many places have strict laws about who can buy or carry weapons, while most regular tools have fewer rules. This is because weapons can easily hurt people if used incorrectly.

4. **Safety Concerns**:
A weapon needs to be handled with more care. If a person is careless, the risk of hurting someone is higher than with normal tools.

## Examples of Basic Weapons

1. **Sticks and Clubs**:
These are among the oldest weapons. A strong stick can be used to hit an enemy. Many ancient people used clubs for defense or hunting.

2. **Spears**:
A spear is like a long stick with a pointed end. It is useful for hunting and for defense because it can keep threats at a distance. Spears were common among many early groups around the globe.

3. **Bows and Arrows**:
A bow and arrows let people attack from a distance. The arrow can be fired at an enemy or an animal without the person getting too close. This gave people a safer way to hunt or fight.

4. **Knives and Swords**:
These are sharp blades used in close combat. A knife is small, easy to carry, and can be hidden if needed. A sword is larger and used for more serious battles. Both rely on their sharp edges to cut or stab.

5. **Slingshots**:
   A slingshot uses a strip of elastic material and a frame to launch small stones. It can be used for hunting small animals and can be made from simple items. Although it seems small, it can still be harmful if used wrongly.

## How Weapons Fit Into Daily Life

Even though many people do not see weapons every day, they still play an important part in our world. Some people keep a weapon for personal safety. In some places, weapons are used by police to keep towns and cities safe. In other areas, soldiers use them at checkpoints.

Hunters use rifles or bows to find animals for food. Farmers in certain regions might need to protect their livestock from wild animals, so they keep a firearm nearby. Sporting events also require weapons, such as archery contests, where participants use bows and arrows to hit targets.

## The Good and Bad of Having Weapons

Weapons can be helpful in preventing threats or in saving lives. For example, someone who lives in a remote area might feel safer if they have something to defend themselves against dangerous animals. Police and soldiers also rely on weapons to protect innocent people from criminals or aggressors.

However, weapons can also create risks. If someone does not store their firearm properly, it might end up in the hands of a child or a thief. This can lead to accidents or crimes. When people have more powerful weapons, conflicts can get worse, and the harm to communities can be larger. Knowing how to handle and store weapons is very important if someone decides to own one.

# How to Tell a Weapon's Effectiveness

Not all weapons have the same power. Some are built for short distances, like swords. Others are built to cover big distances, like rifles and cannons. There are a few factors that can help us understand how effective a weapon might be:

1. **Range**:
   This is how far a weapon can reach to harm a target. Spears have a longer range than knives, while guns can reach farther than spears.
2. **Damage**:
   This is how much harm a weapon can cause. It depends on the weapon's design. For instance, a sword can cut deeply, but it might not work well against someone very far away.
3. **Speed**:
   A fast bullet or arrow makes it harder for the target to move out of the way. Speed often ties to the design and the power behind launching a projectile, like gunpowder in a firearm.
4. **Ease of Use**:
   A weapon is more effective if a person can learn to use it quickly and handle it safely. A complicated device might do more damage but takes longer to master.
5. **Reliability**:
   A weapon that works in different weather and doesn't break easily is more reliable. If it jams or breaks at a bad time, it becomes useless.

---

# Everyday Tools vs. Improvised Weapons

Sometimes people use everyday tools in ways they were not meant to be used. A baseball bat can be a sports item, but it can become a weapon if used to hit someone. A sharp gardening tool could also cause harm. These examples are usually called "improvised weapons" because they are not meant to be used for fighting, but can be if someone chooses to do so.

What makes something an improvised weapon is the purpose the user has in mind. Even a belt can be used to tie someone's hands or swing at them. This is one reason that laws about weapons can be tricky. Many normal objects can be used in a harmful way, yet they are not banned because they have harmless uses as well.

## When Weapons Become a Concern

1. **Accidents**:
   If someone leaves a loaded gun on a table, a child might pick it up. This can cause terrible harm. Keeping weapons unloaded and locked in a safe place is important.
2. **Unauthorized Use**:
   People who should not have a weapon might steal it or buy it illegally. If a dangerous person gets hold of a powerful weapon, they can harm many people.
3. **Escalation of Violence**:
   If two groups have an argument but neither has a weapon, they might end up fighting with fists or deciding not to fight at all. If both have guns, the risk of serious harm is higher. Thus, weapons can make conflicts more severe.
4. **Fear in Communities**:
   If many people carry weapons openly, it can make others worried. Some might feel safer, but others might feel uneasy, leading to tension in a neighborhood.

## Questions People Have About Weapons

- **Who should have weapons?** Some argue that only certain people, like police or trained professionals, should have them. Others believe people have a right to defend themselves.

- **How do we control them?** Different places have different laws. Some require licenses, background checks, or waiting periods before buying a weapon.
- **Why do countries keep making them?** Countries often want to protect themselves or keep up with other countries. This leads to what some people call an "arms race," where each side tries to have better weapons than the other.

## Weapons and Technology

As technology improves, so do weapons. Today, some weapons use advanced electronics, lasers, or even computers to guide them. What began as simple sticks and stones has grown into a wide range of items capable of big destruction. This is why many people try to keep track of weapon development and limit how powerful weapons can get.

On the other hand, simpler weapons still exist. People in rural areas might still use bows, arrows, or knives, either because they like the tradition of using them or because they cannot afford modern weapons.

## The Role of Training

Knowing how to handle a weapon properly is key. Without training, a person might hurt themselves or someone else by mistake. Training includes learning how to aim, fire, and maintain the weapon. It also involves understanding safety rules, like checking if a gun is loaded or pointing it in a safe direction.

Training also covers the moral and legal parts of using a weapon. When is it right to use it? In many places, a person can only use a weapon if there is a real threat. Firing a gun for fun in a crowded area, for example, is dangerous and often illegal.

# CHAPTER 2

## A BRIEF LOOK AT THE HISTORY OF WEAPONS

Weapons have been around as long as humans have. In fact, some experts say the first tools used by early humans were actually weapons, such as stones used to smash objects or sticks used to protect against animals. Over thousands of years, weapons developed from simple shapes to more refined tools, such as spears, swords, and eventually firearms. In this chapter, we will see how weapons changed over time, from the earliest days up to some of the most important inventions in history. We will also explore how new ideas and materials pushed weapon development, leading us to the modern devices we see today.

### Early Tools Become Weapons

When humans first started walking the earth, they faced many threats: wild animals, other groups of people, and the challenges of nature. They soon discovered that they could use certain objects to give themselves an edge. A hard rock could break open nuts but also hurt an enemy. A strong stick could help someone get fruit from a tall branch but also strike something dangerous.

Over time, early humans realized that if they sharpened a rock, it could become more useful for hunting and defending. This was likely the first step toward making specialized weapons. People began to shape stones into arrowheads or to fit them on the tips of sticks. This allowed hunters to kill animals from a small distance rather than getting too close.

## Spears and the Advantage of Distance

One of the biggest early breakthroughs was the spear. A spear has a pointed tip, often made of a stone or a piece of bone, attached to a wooden shaft. This design gave people an advantage because they did not have to stand right next to a threat. They could throw the spear or thrust it from a short distance. For hunting, this meant they could stay back while aiming at a large animal. For fighting, it meant they had a better defense against opponents who only had knives or clubs.

Spears were also important in forming early armies. Groups that had well-made spears and knew how to use them in a formation could protect themselves against attacks. This idea of teamwork, where people stand in lines or groups with their weapons pointed outward, became a central part of organized warfare in many ancient societies.

## The Rise of Bows and Arrows

At some point, people invented the bow and arrow. This might have happened in different parts of the world at different times, but the idea was the same: a curved piece of wood with a string that could launch a lightweight projectile at higher speed and distance. Arrows could be made with stone or bone tips, and fletching (feathers) at the back end to help with flight stability.

The bow and arrow changed hunting and fighting. Instead of getting close to a wild animal or an enemy, people could strike from even farther away than with a spear. This improved safety for the archer. Over time, more advanced bows were developed, like the composite bow, which used different materials such as wood, horn, and sinew. This made bows stronger and more flexible.

# The Age of Metal

A major leap in weapon history happened when humans learned to melt and shape metals. First came bronze, a mix of copper and tin. Bronze swords, knives, and spear tips were stronger and more durable than stone or bone. This gave groups with metal weapons a big advantage in conflicts.

Later, people learned how to make iron weapons. Iron, and especially steel (which is iron combined with carbon), was even tougher than bronze. This led to better swords, stronger armor, and more effective spear tips. In regions like Africa, Asia, and Europe, metal weapons helped form powerful kingdoms and empires. Warriors with iron or steel equipment often defeated those who still relied on older tools.

# Advances in Swords and Armor

During the iron and steel periods, swords became more refined. They were shaped in different ways to match different fighting styles. Some swords were short and good for thrusting. Others were long and curved for slicing. Each culture developed its own style based on local resources and fighting methods. For example:

- **Short Swords**: Popular among Roman soldiers, who fought in tight formations.
- **Long Swords**: Used by knights in medieval Europe, who relied on heavy swings and strong armor.
- **Curved Swords**: Seen in places like the Middle East, where riders on horseback needed a quick, slashing weapon.

Armors also developed. Leather armor, then chain mail, and later steel plates were used to protect the body. This arms race between weapon strength and armor toughness led to constant changes in design.

# The Impact of Gunpowder

A massive turning point in the history of weapons was the discovery and use of gunpowder. Gunpowder was first made in ancient China. People noticed that a mix of saltpeter, sulfur, and charcoal could create a powerful explosion when lit. At first, it was used for fireworks and signals. Soon, though, people tried using it to power projectiles.

The first firearms were crude and not very accurate, but even a simple cannon had a huge effect on how wars were fought. It could break down castle walls and scare armies that had never seen or heard anything like it. Over time, muskets and rifles replaced bows and arrows. The reasons for this were:

1. **Power**: Gunpowder weapons could penetrate armor more easily.
2. **Range**: Cannons could hit targets far away.
3. **Fear Factor**: Loud booms and smoke made them intimidating.

Though early firearms were slow to reload and prone to misfiring, technology improved until they became the main weapons for soldiers worldwide.

---

# Muskets, Rifles, and Modern Firearms

Muskets were one of the first handheld firearms used by many armies. They had smooth barrels and were fired by lighting the gunpowder inside with a spark or a match. Reloading took time, so soldiers stood in lines and fired in volleys before reloading again.

Later, rifling—grooves cut inside the barrel—helped bullets spin, improving their accuracy. These weapons were called rifles. They changed battles by allowing soldiers to hit targets at longer distances.

Repeating rifles and revolvers added new layers to warfare. Soldiers no longer had to reload after every shot. They could fire multiple rounds before needing to stop. This increased the speed and frequency of shots in battle, giving a strong advantage to armies that had them.

## The Industrial Period and Mass Production

When factories became common, weapons could be made in large numbers. Before that, a blacksmith made one sword at a time. Now machines could make many guns or swords in a single day. This meant countries with factories could arm their soldiers faster and cheaper. It also made weapons more uniform in design and quality.

The Industrial Period also saw changes in how wars were fought. Armies grew bigger, and they carried standardized weapons. Supply lines became more important, as governments had to move weapons to wherever the battles were. We will talk more about supply and distribution in later chapters, but it is enough to say here that mass production changed the face of conflict forever.

## Key Weapon Inventions in History

1. **The Catapult**:
   A device using tension or torsion to launch stones or other projectiles over walls. Widely used in ancient and medieval times.
2. **The Longbow**:
   A tall bow that offered more power than a regular bow. It could shoot arrows farther and with greater force, giving English archers a great advantage in battles.

3. **Gunpowder Cannon**:
   Early cannons changed city defenses forever. High walls that once seemed impossible to breach could be knocked down with enough cannon fire.
4. **The Machine Gun**:
   Invented in the late 19th century, it could fire many bullets quickly without reloading after each shot. This caused huge changes in how battles were fought and is considered one of the deadliest shifts in warfare tactics.

---

## The Move to Modern Times

As time went on, weapons got more complex. By the 20th century, we had airplanes that could drop bombs, submarines with torpedoes, and tanks with powerful cannons. With each new invention, armies tried to outdo one another, leading to global conflicts like the World Wars. During these wars, technology advanced even faster, resulting in deadlier and more precise weapons.

Eventually, atomic bombs were created, which showed a new level of destruction. After the first use of an atomic bomb, many nations started building their own nuclear weapons, worried about being left behind. We will talk more about these in Chapter 10.

---

## How Ancient Inventions Still Affect Us

Many old inventions, like the bow, spear, and sword, are still used today in some form. Some people practice historical martial arts. Others use bows for hunting or archery competitions. Swords may not be common in modern warfare, but they are part of ceremonial uniforms and are often kept as collectibles.

Old ideas also inform new designs. The concept of launching a projectile—like a rock from a catapult—lives on in missiles and rockets, though the technology is much more advanced. The idea of harnessing explosive power from gunpowder has expanded into all sorts of modern weapons, from grenades to advanced artillery.

---

## Lessons Learned from the Past

1. **Innovation**:
   New materials and ideas can change the world. Bronze, iron, and steel all shaped entire eras of history.
2. **Balance of Power**:
   Those with stronger weapons often had more control. This led to empires rising and falling. It also caused arms races where different groups tried to keep up or surpass each other.
3. **Impact on Society**:
   Whenever a new weapon became common, ways of life changed. Castles stopped being so strong once cannons appeared. Swords became less useful once guns were easy to load. Each shift in weapons forced societies to adapt.
4. **Moral Questions**:
   From the beginning, people wondered if certain weapons were too harmful. This question continues today, especially when it comes to very destructive devices.

---

# CHAPTER 3

## HOW MODERN WEAPONS ARE DESIGNED

Modern weapons are created through a careful process that blends science, engineering, and planning. Designers think about many details before they build anything, like what the weapon should do, who will use it, and how it will perform in different conditions. This chapter will explain how weapon designs move from ideas to real objects. We will look at the steps involved, the new technologies used, and the reasons why testing and improvement never stop.

## Starting with a Need or a Problem

The first step in designing a modern weapon is to figure out why it is needed. Often, a group like an army or a police force will say, "We need a weapon that can do this specific job." For instance, a country may want a new rifle that fires more accurately or a tank that can move faster on rough ground. Sometimes, weapon makers themselves come up with ideas they think will be useful or profitable.

1. **Defining the Purpose**: Is this weapon meant to stop enemies at a distance? Is it meant to protect a base? Will it be carried by a soldier, or placed on a vehicle? Answering questions like these helps designers know the shape, size, and features the new weapon should have.
2. **Setting Requirements**: Once the purpose is clear, the team sets goals. For example:
    - How far should it shoot?
    - How many rounds can it fire before reloading?
    - How heavy or light should it be?
    - How accurate must it be?

These requirements serve as a guide for everyone working on the project.

## Gathering Ideas and Drawing Plans

After the purpose is decided, engineers and designers start drafting plans. They may use pencils, paper, and computer software called CAD (Computer-Aided Design). CAD programs allow them to build virtual models of the weapon, showing how each part fits together. This helps them spot problems before they spend money on making real parts.

- **Brainstorming**: Team members offer ideas based on their different backgrounds. Mechanical engineers might suggest ways to reduce weight. Materials experts might recommend stronger metals or plastics.
- **Rough Sketches**: They create rough images to visualize the main shape and size. They might draw how a barrel lines up with a firing mechanism, or how a handle fits a soldier's grip.
- **Detailed Blueprints**: As the design becomes more certain, the team forms detailed blueprints, showing every bolt, spring, and surface measurement.

## Choosing Materials

Deciding which materials to use is a major part of modern weapon design. Different parts of a weapon might need different materials. For example, the barrel of a rifle must handle a lot of heat and pressure, so it might be made of steel or special alloys. A hand grip might be plastic or composite material to make it lighter and easier to hold.

1. **Strength**: Weapons must handle shock and stress. A rifle barrel has to endure the force of repeated shots. A tank's armor must stop incoming fire.
2. **Weight**: Soldiers do not want heavy equipment that slows them down. Lighter materials make weapons easier to carry, but they must still be strong enough.
3. **Heat Resistance**: Many weapons get hot when fired repeatedly. The right metals or composite materials can resist high temperatures without breaking.
4. **Cost**: Some materials are very good but too expensive. Designers must balance performance with what buyers can afford.

## Building a Prototype

Once the design is on paper and the materials are chosen, the team builds a prototype. A prototype is a first version of the weapon that can be tested in real life. Building prototypes can be expensive, but it helps the team learn what works and what does not.

- **3D Printing**: Nowadays, parts of prototypes might be 3D printed. This allows designers to quickly create shapes for testing without waiting for a factory to make them.
- **Handmade Parts**: Some parts, like barrels or triggers, still need skilled workers to shape them precisely. Even with modern machines, careful craftsmanship can matter a lot.
- **Testing Fit and Comfort**: Engineers might give the prototype to a soldier or a tester to see if it is comfortable to hold and easy to aim.

## Testing and Evaluation

The next step is to test the prototype. Testing is extremely important because weapons must be reliable. No one wants a weapon that jams during a critical moment.

1. **Accuracy Tests**: Firing many rounds to see how close they land to the target. This can be done at indoor ranges or outdoor fields.
2. **Durability Tests**: Seeing if the weapon still works after being dropped, exposed to rain, dust, cold, or extreme heat.
3. **Rate of Fire Tests**: For automatic weapons, testers check how many bullets can be fired in a minute. They also watch for overheating.
4. **Stress Tests**: They push the weapon harder than normal use to see if it breaks. This might include firing more rounds than usual or using the weapon in tough weather.
5. **Ergonomics Checks**: Making sure the shape is comfortable for a person to use for a long period. If it is too heavy, or the recoil is too strong, changes may be needed.

When a problem shows up, the design team notes it and returns to the workshop. They adjust the weapon, swap out parts, or change the materials. Then they test it again. This cycle can happen many times until the designers are happy with the results.

---

## Safety Considerations

A good weapon design also includes safety features. For example, a gun might have a safety switch that stops it from firing by accident. The design might include clear indicators to show if it is loaded. For bigger weapons like tanks or aircraft systems, designers must ensure that soldiers can operate them without harming themselves.

Even well-designed weapons can be dangerous if used incorrectly. That is why training and proper instructions are crucial. But from the designer's view, they must do everything they can to lower the chance of accidents or misuse.

---

## The Role of Computers and Simulation

Computer simulation is a modern tool that helps save time and cost. Before building a real prototype, engineers can run tests inside a computer program to check how a weapon might behave under different conditions.

- **Heat and Stress Simulation**: Computers can predict how hot the barrel will get after many shots. They can show where metal might crack or bend.
- **Trajectory Simulation**: For weapons that fire projectiles, simulations can predict the path of a bullet or shell in various wind and weather conditions.
- **Blast and Shock Effects**: If the weapon uses explosives, simulations can tell how strong the blast might be and how far it could reach.

While simulations do not replace real-world tests, they give designers valuable hints on where problems might occur, reducing the time spent fixing issues after a physical prototype is made.

---

## Collaboration with Different Experts

Designing a modern weapon is rarely the work of just one person. Many experts come together:

1. **Engineers**: They handle the mechanical parts, electronics, and software.
2. **Materials Scientists**: They research and choose metals, plastics, and composites.
3. **Ballistics Experts**: They specialize in how projectiles fly, ensuring the weapon hits its target accurately.
4. **Human Factors Specialists**: They make sure the weapon is comfortable and easy to use.

5. **Budget and Finance Teams**: They keep track of costs and make sure the project stays within limits.
6. **Legal Advisors**: They ensure the design meets government rules and does not break any regulations.

This teamwork is necessary because modern weapons can be very complex.

## Innovations in Modern Weapon Design

Technology has greatly changed how weapons are made:

- **Smart Components**: Some weapons have computer chips that gather data about targets or help with aiming.
- **Modular Designs**: Designers make weapons with pieces that can be swapped out, allowing the user to change barrels, scopes, or grips for different tasks.
- **Lightweight Materials**: Carbon fiber and high-strength polymers reduce weight while staying strong.
- **Improved Sights and Optics**: New scopes, night-vision devices, and laser range finders help shooters see targets clearly in various conditions.

These features can make a big difference in modern combat or law enforcement, where speed and accuracy matter a lot.

## Balancing Performance with Ethics

As weapons become more advanced, people worry about misuse or harm to bystanders. Designers sometimes face tough questions: Should they add features that make it harder for unauthorized users to fire a weapon? Should they build weapons that track location data, so they are harder to steal?

While the main job of engineers is to meet the requests of those who order weapons, there is also a sense of responsibility. Some companies add extra safety features, even if they are not required. Others work with governments to follow strict export rules or user guidelines.

## Production and Assembly

When testing shows the design meets the goals, the next step is mass production:

1. **Factory Set-Up**: Machines, assembly lines, and workers are arranged to make large numbers of the weapon quickly.
2. **Quality Control**: Each unit is checked to ensure it meets standards, such as correct barrel length or proper firing function.
3. **Packaging and Delivery**: Finished weapons are packed and shipped to those who ordered them, like military bases or police departments.

During production, small changes might still happen if workers find an easier way to assemble parts, or if a minor fault appears that needs fixing.

## Upgrades and Future Designs

Weapons are not finished once they are produced. Over time, new discoveries or problems might arise. Some weapons get updated with better electronics or improved barrels. This leads to "versions" or "marks" of a weapon, such as Mark I, Mark II, and so on. Each version might address previous concerns or add new features.

In some cases, a completely new design is needed. If technology advances a lot or if the demands of the battlefield change, a weapon might become outdated. Then, designers start again, studying lessons from the old design and aiming for something better suited to new challenges.

## Conclusion of Chapter 3

Modern weapon design is a detailed process that blends creativity and scientific rigor. It starts with a clear need, moves through planning and prototyping, and involves many rounds of testing to ensure that the final product is reliable. Advanced computers, new materials, and teamwork across different fields all play a part in bringing these ideas to life.

Even once a weapon is in service, work does not stop. Designers and engineers keep seeking ways to make weapons safer, more accurate, lighter, or more powerful—while also thinking about costs and ethical concerns. This constant effort shapes how militaries and other groups around the world keep up with changing needs.

# CHAPTER 4

## SMALL ARMS AND THEIR ROLE

Small arms are weapons that can be carried and used by one person. They include pistols, rifles, shotguns, and submachine guns. Unlike heavy cannons or missiles, small arms are often meant for closer combat or personal defense. Yet, they play a major role in many armed forces, police units, and private settings. In this chapter, we will talk about the different types of small arms, how they work, and where they are used.

## What Are Small Arms?

Small arms are generally handheld firearms or even simpler weapons that can be operated by an individual without needing extra equipment. Common examples are:

1. **Handguns (Pistols and Revolvers)**: Short-barreled firearms easy to carry and conceal.
2. **Rifles**: Long-barreled firearms designed for accuracy at medium to long ranges.
3. **Shotguns**: Firearms that usually fire a spread of small pellets (shot) or a single slug.
4. **Submachine Guns**: Fully automatic or semi-automatic weapons that fire pistol cartridges.

Though these weapons are smaller compared to artillery or tanks, they can still be very dangerous. They come in many shapes and sizes, each made for a specific task.

## Pistols and Revolvers

**Pistols** and **revolvers** are sidearms typically used at short range. They are designed to be carried easily in a holster. Here is how they differ:

- **Pistol (Semi-automatic)**: Uses a magazine to feed bullets into the chamber. After each shot, the weapon automatically loads the next bullet.
- **Revolver**: Has a revolving cylinder that holds a fixed number of bullets. Each time you fire, the cylinder turns to line up the next round.

Both are popular with police, military officers, and civilians who are allowed by local laws to own a handgun for protection or sports shooting. Their small size makes them handy, but they have less power and range than rifles.

---

## Rifles

Rifles are long guns with spiraled grooves inside the barrel called **rifling**. This rifling makes the bullet spin, increasing its accuracy over longer distances. Some rifles are designed for hunting, others for military use, and some for sports competitions. They can be **bolt-action**, **lever-action**, or **semi-automatic**, depending on how they load the next round.

- **Bolt-Action**: The user lifts and pulls back a bolt to eject the empty shell and push a new bullet into the chamber. It is common in hunting and sniper rifles because of its simple, accurate design.
- **Lever-Action**: Found in some older-style rifles, a lever near the trigger is pumped forward and back to load the next round.

- **Semi-Automatic**: The rifle uses the energy from firing a shot to load the next round automatically. The user just pulls the trigger once for each shot.

Military assault rifles, such as those used by soldiers, can switch between semi-automatic and burst or full-auto modes. These rifles are designed for combat, with shorter barrels and more rugged construction than many hunting rifles.

## Shotguns

A **shotgun** usually has a smooth barrel that fires many small pellets at once. This is effective for hitting moving targets, like birds or running animals when hunting. It is also used by police and military units for close combat or breaking open doors.

- **Pump-Action**: After each shot, the user slides a pump handle to eject the empty shell and load a fresh one.
- **Semi-Automatic**: Works like a semi-automatic rifle, automatically loading the next shell from a magazine tube.
- **Break-Action**: The barrel and breech fold open for manual loading of shells. This style is often used in sports shooting.

Because the shot spreads out, precise aiming is less critical at short ranges, but the spread also means the pellets lose power as they travel further.

## Submachine Guns

A **submachine gun (SMG)** is a small automatic weapon that usually fires pistol-caliber bullets, like 9mm. They are meant for close-range fights, where their rapid-fire ability can be an advantage. Think of them as a middle ground between a pistol and a rifle. They are often used by special forces, vehicle crews, or security teams in tight spaces. SMGs are lighter and shorter than most rifles, making them easier to move around in urban or indoor environments.

## Uses of Small Arms

Small arms are everywhere, from police stations to private homes. Here are some of their major uses:

1. **Military**: Soldiers carry rifles or sidearms. Special units may have submachine guns for specific tasks, like boarding enemy ships or fighting in narrow buildings.
2. **Law Enforcement**: Police often use handguns and sometimes shotguns or rifles for tougher operations.
3. **Security Guards**: Guards in some countries carry pistols for protection against theft or attacks.
4. **Hunting**: Rifles and shotguns are common for hunting animals. The type used depends on the animal's size and the hunting laws in the area.
5. **Sports Shooting**: People compete in target shooting events using pistols, rifles, or shotguns to test accuracy and skill.
6. **Personal Defense**: In places where law allows it, some individuals keep small arms at home or on their person for self-defense.

## How Small Arms Work

Most small arms follow a similar cycle when firing:

1. **Loading**: A bullet or shell is placed into the weapon. This can be done by inserting a magazine in a pistol, chambering a round in a rifle, or placing shells in a shotgun's tube or breech.

2. **Firing**: When the user pulls the trigger, the firing pin hits the primer at the base of the bullet or shell, igniting the propellant (gunpowder).
3. **Bullet Travel**: The burning propellant creates gas pressure that pushes the bullet out of the barrel. In a shotgun, it pushes out pellets.
4. **Ejection and Reloading**: If it is a repeating firearm, the empty case is ejected, and a new round is moved into position for the next shot.

In some weapons, steps 3 and 4 happen automatically. In others, the user must work a bolt or pump a handle. Understanding these steps is crucial for safe handling and maintenance.

---

## The Importance of Maintenance

A small arm is only reliable if it is well-maintained:

- **Cleaning**: After firing, residue from the burning propellant can build up in the barrel and other parts. Regular cleaning prevents rust and removes dirt that can cause jams.
- **Lubrication**: Moving parts need a thin layer of oil to reduce friction. Too much oil can attract dust, so a balance is needed.
- **Inspection**: Checking parts for wear or damage helps avoid malfunction. A cracked barrel or worn firing pin can be very dangerous.

Many accidents happen when people do not keep their weapons clean or fail to fix a damaged part. Proper maintenance can lengthen a weapon's life and keep it working safely.

---

## Safety Features

Small arms often include several safety features to reduce accidents:

- **Manual Safety Switch**: A lever or button that physically stops the weapon from firing. The user must switch it off before pulling the trigger.
- **Trigger Safety**: Some pistols have a small safety built into the trigger. The shooter must place their finger properly for the weapon to fire.
- **Drop Safety**: Designed so that if the weapon is dropped, it will not discharge accidentally.
- **Loaded Chamber Indicator**: A small tab or marking that shows if a round is in the chamber.

Even with these features, a user must handle small arms with care. Safety depends on proper training, discipline, and following rules like never pointing the weapon at anything you do not want to harm.

---

## The Spread of Small Arms Around the World

Small arms are the most common type of weapon globally. They are relatively cheap and easy to transport. This means they can end up in many places, including spots they should not be. Many conflicts around the world involve these weapons, and controlling their spread can be tough. International organizations and governments sometimes try to limit who can buy or ship them, but illegal sales happen in some areas.

This wide availability can lead to problems, such as violence in regions where many small arms circulate with no oversight. That is why many groups call for stricter controls or better tracking of sales.

# Technology in Small Arms

Just like larger weapons, small arms also benefit from new technology:

1. **Smart Guns**: Some designs include fingerprint or digital locks, so only approved users can fire them. These are not yet widespread, but they are an idea meant to reduce misuse.
2. **Improved Ammunition**: New bullet designs can fly straighter or have less recoil. Special coatings can also reduce barrel wear.
3. **Optics and Accessories**: Red dot sights, telescopic sights, and night-vision scopes help shooters see targets in low light or at a distance.
4. **Modular Systems**: Some rifles allow different barrels, stocks, and attachments to be swapped, changing caliber or function quickly.

While these upgrades can make small arms more efficient and safer, they can also raise worries if such features become common in illegal markets.

---

# Training and Responsibility

Owning or using a small arm requires knowledge and training. A new user must learn:

- **Safe Handling**: How to hold the weapon correctly, load it properly, and keep the muzzle pointed away from others.
- **Aiming and Firing**: Controlling breathing, posture, and trigger pull for accurate shots.
- **Storage**: Keeping the weapon locked away so children or unauthorized people cannot access it.

- **Legal Duties**: Understanding local laws about carrying, transporting, or using firearms.

In many places, there are tests and background checks before someone can buy a small arm. These rules try to keep firearms out of the hands of those who might use them to hurt others.

## Balancing Power and Control

Small arms can bring protection to individuals and communities, but they also pose risks. If too many people carry them without proper training or oversight, accidents and crimes may go up. Finding the right balance between allowing ownership for defense and making sure they do not end up with criminals is a challenge for governments worldwide.

Different countries handle this in different ways. Some have strict laws, requiring background checks, licenses, and waiting periods. Others have fewer rules, letting people buy firearms more freely. Police, lawmakers, and citizens often debate which approach keeps everyone safest.

## Future Trends for Small Arms

Small arms are always evolving. Designers are looking into lighter materials, better electronics, and ways to personalize guns to each user. Some ideas aim to reduce recoil, improve accuracy, or even track how many shots have been fired. There are also ongoing debates about "smart gun" technology, which might prevent anyone but the owner from firing the weapon.

Still, basic principles like safety, responsibility, and legality will remain important. No matter how advanced small arms become, they will keep playing a key role in personal defense, law enforcement, and military tasks around the world.

# CHAPTER 5

## LARGE-SCALE GUNS AND CANNONS

Large-scale guns and cannons are powerful artillery pieces capable of firing shells or heavy projectiles over long distances. Unlike small arms, which are carried by a single person, these bigger weapons need many people to operate them or a special mount, such as a vehicle or a stationary platform. They can launch projectiles that cause significant damage to structures and vehicles, and they are a mainstay in most military forces around the globe. In this chapter, we will look at the history of cannons, the types of large-scale guns, how they are built, and how they are used in modern times.

## Early History of Cannons

Long ago, cannons were not as refined as they are now. The first cannons came from the discovery of gunpowder in China. People learned they could use the explosive force of gunpowder to push a heavy stone or metal ball through a tube. Early cannons were made from materials like cast iron or bronze. They were often placed on wooden carriages with wheels, so they could be moved around the battlefield.

However, these early cannons were crude. They were hard to aim, took a long time to load, and sometimes blew up if the metal or the design was poor. Still, they gave armies a big advantage against enemies who lacked similar firepower. Cannons could smash through castle walls, changing how battles were fought. Defenders had to rethink fortifications, leading to new designs with thicker walls and angled edges.

As time went on, artisans and metalworkers improved the making of cannon barrels. By refining the shape and using better metal, the cannons became less likely to explode by accident. Commanders realized that these weapons could decide the outcome of sieges. An army with many cannons could force a city to surrender by battering its walls from a safe distance. This shifted the power balance, giving those with advanced artillery an edge.

## The Move from Muzzle-Loading to Breech-Loading

Early cannons were **muzzle-loading**, which means you loaded gunpowder and a ball from the front (muzzle) of the barrel. This was time-consuming. Soldiers had to swab out the barrel to clean hot debris, then carefully pour gunpowder, add wadding (a piece of material to hold the powder in place), and finally push in the cannonball. If they were not careful, a stray spark could ignite the powder too soon.

Eventually, weapons makers developed **breech-loading** cannons. Breech-loading means the rear (breech) of the barrel can open so the shell and charge can be placed quickly. This made firing faster and safer. Breech-loading cannons also allowed for better sealing of the gases created by the burning powder, giving the projectile more force and improving accuracy.

## Types of Large-Scale Guns

There are different kinds of big guns, each designed for a special role:

1. **Field Artillery**
   These are guns used on the battlefield to support infantry (foot soldiers). They often have wheels and can be towed by vehicles or horses. Field artillery includes howitzers, which

fire projectiles in a high arc, and guns that fire at a flatter angle.
2. **Siege Artillery**
Siege guns are larger than field artillery and are used mainly to break down fortifications like walls and bunkers. They may be less mobile but pack a strong punch. Historically, armies had to bring huge cannons to battle when trying to conquer well-defended castles or cities.
3. **Coastal Artillery**
These cannons defend coastlines from enemy ships. Positioned in forts near the water, they can fire at vessels trying to come close. Modern coastal batteries might include guided missiles, but in the past, large cannons kept warships away.
4. **Naval Guns**
Warships carry big guns that are mounted in turrets. These guns can rotate to aim at targets in different directions. Their size can vary from smaller-caliber guns to massive, battleship-class guns that can hit targets many miles away.
5. **Anti-Aircraft Guns**
These guns are aimed at the sky to shoot down planes or other flying objects. They fire shells that may explode near targets, sending shrapnel in all directions. Modern anti-air systems often use missiles, but some still rely on rapid-firing cannons.

---

# Artillery Shells and Ammunition

Large-scale guns do not just fire simple cannonballs anymore. Today, they shoot shells that can have different uses:

- **High-Explosive (HE) Shells**: These shells explode upon hitting a target or when their built-in timer ends. They cause damage with the blast and flying fragments.

- **Armor-Piercing (AP) Shells**: These shells have a hard metal tip to penetrate thick armor. They are often used against tanks and fortifications.
- **Smoke Shells**: These do not cause harm directly but produce a thick cloud of smoke to hide troop movements or block an enemy's view.
- **Illumination Shells**: These contain bright flares that light up the battlefield at night so soldiers can see what they are doing.
- **Chemical or Specialized Shells**: Certain shells might carry special materials for specific missions. However, many are restricted or banned by international rules.

Artillery shells are larger and heavier than small arms ammunition. They may require multiple people or machines to handle them safely. The shell's fuse, which is the device that triggers the explosion, can be set to detonate on impact or above the ground to rain fragments over a wider area.

## How Cannons Work

A cannon operates similarly to a smaller firearm, but on a larger scale. When the shell is loaded into the barrel and the powder is ignited, expanding gases push the shell forward. Recoil forces the gun backward, so cannons often have recoil systems that help absorb this backward push. Springs, hydraulic cylinders, or other devices keep the artillery piece steady and allow it to return to its original firing position.

Aim is controlled by adjusting elevation (up and down) and traverse (left and right). Larger cannons might need a crew that includes a gunner, loader, and others. Modern systems often include computers that calculate the angle and power needed to hit a target at a certain distance, factoring in wind, weather, and the type of ammunition.

# Modern Artillery Technology

Technology changed artillery in big ways:

1. **Self-Propelled Artillery**: Rather than towing a cannon, many armies use self-propelled guns. They resemble tanks with large cannons mounted on top. They can move quickly, fire, and then relocate to avoid counterattacks.
2. **Precision-Guided Shells**: Some modern shells have guidance systems similar to missiles. These shells can adjust their path mid-flight to strike a target with high accuracy. This reduces the number of shots needed and lowers the risk of hitting the wrong place.
3. **Automated Loading**: Advanced artillery vehicles might load shells automatically, reducing the workload on the crew. This can also speed up how fast the gun fires.
4. **Digital Fire Control**: Computers calculate exact firing solutions. A soldier inputs the target's location, and the system sets the correct elevation and direction. This helps land shots more accurately on the first try.
5. **Radar and Detection**: New radar systems track the path of shells. By watching the arc of an incoming round, an army can figure out where it was fired from and quickly return fire.

All these advances make modern artillery quicker, more precise, and harder to locate. However, they also increase complexity and cost. Crews need training in how to use electronic fire-control systems and maintain more delicate parts.

---

# Field Deployment and Tactics

Large-scale guns are used in many ways on the battlefield:

- **Support for Infantry**: Artillery can fire ahead of advancing troops, hitting enemy positions. Once the artillery finishes, the infantry moves in. This is called a "fire and maneuver" strategy.
- **Defensive Fire**: If enemy forces attack, artillery can lay down a barrage of shells to slow or stop them before they get too close.
- **Counter-Battery Fire**: When the enemy fires artillery, radar or other detection tools can find their location. The defending side then returns fire to destroy the enemy's guns.
- **Psychological Effect**: The loud sound and impact of artillery can lower enemy morale. On the flip side, it can also put stress on friendly troops if not coordinated properly.

One challenge is avoiding harm to civilians or allied forces. Because artillery shells can travel far, there is a risk of hitting unintended targets if the data used for aiming is wrong. Crews must follow strict procedures to confirm coordinates before firing.

## Maintenance and Logistics

Keeping large-scale guns in good condition can be a challenge. Their barrels wear down due to high heat and pressure, so they must be replaced after firing a certain number of shells. Dirt, sand, or rust can also cause problems if not cleaned properly. Crews spend time checking seals, lubricating moving parts, and adjusting recoil systems to keep the weapon reliable.

Logistics matter too. Artillery uses heavy shells that need trucks or other vehicles to transport them to the front lines. A single gun might need hundreds of shells to support a day of fighting. Storing and moving these shells safely requires planning. If a supply chain is broken, the artillery piece becomes less useful because it cannot fire without ammunition.

## Notable Examples in History

- **"Big Bertha"**: A famous German siege howitzer used during World War I. It could launch huge shells over several miles, wrecking enemy forts.
- **Paris Gun**: Another World War I German creation, it was designed to fire shells that soared high into the atmosphere, reaching Paris from very far away.
- **M1 Howitzer (United States)**: A well-known field gun from World War II, prized for its reliability.
- **Modern Self-Propelled Guns**: Many countries now have self-propelled artillery vehicles like the M109 (U.S.), PzH 2000 (Germany), and others. These combine mobility, armor, and firepower in one package.

These examples show how artillery can shape battle outcomes, from old-style siege warfare to fast-paced modern conflicts.

---

## Civilian and Non-Military Uses

Cannons and large-scale guns are mostly tied to warfare, but some have had non-military uses. For instance, special cannons test how aircraft materials handle bird strikes, by firing bird-like objects at airplane windows or engines. There are also devices that launch things like fire-retardant chemicals to fight wildfires, though these are not usually referred to as "cannons" in the same sense as military weapons.

Mostly, though, the big guns remain in the hands of militaries. Their strong destructive power makes them unsuitable for everyday use. People do see historic cannons on display in museums or used during ceremonies and reenactments. These older cannons often do not fire live rounds anymore, but they remind us of how important artillery was in shaping history.

## Challenges and Concerns

Large-scale guns can cause widespread damage if used without caution. When shells explode, they can harm people and destroy buildings. In areas with many civilians, there is a risk of collateral damage. Some organizations want stricter rules on the use of heavy artillery, especially in densely populated places.

There is also the problem of older artillery being sold or passed along in illegal arms deals, ending up in places with unstable governments. Such weapons can increase the danger in conflict zones. Efforts exist to keep track of these weapons, but they are not always successful.

## Future Trends

Artillery continues to evolve. Developers are exploring railguns, which use electromagnetic force rather than gunpowder to fire projectiles. These can reach extremely high speeds and longer distances. Others are looking at advanced rocket-assisted shells that can travel farther while staying accurate.

Also, modern data tools make it simpler for artillery crews to plan their shots. Mobile devices can share target details across an entire network of units, so multiple guns can focus on one location at the same time. As with many weapon systems, the trend is toward greater precision, automation, and integration with digital networks.

# CHAPTER 6

## TANKS AND ARMORED VEHICLES

Tanks and other armored vehicles are moving machines covered in thick protective layers, carrying powerful weapons to support troops or break through enemy defenses. They stand out in battle for their ability to move over rough ground, resist small arms fire, and deliver strong attacks with their main guns. In this chapter, we will look at how tanks and armored vehicles were developed, what they do on the battlefield, how they are designed, and how they continue to change in modern warfare.

## Early Development of Tanks

The idea of a heavily protected vehicle that could roll across enemy lines began in World War I. At that time, both sides were stuck in trench warfare, and soldiers needed a way to cross barbed wire and trenches without getting hit by machine-gun fire. Engineers thought of building a large, metal machine on tracks that could carry soldiers safely and destroy obstacles.

The British were among the first to create these vehicles, calling them "tanks" to hide their true purpose. Early tanks were slow, noisy, and often broke down. Still, they had a significant effect when they worked. They could roll over trenches, crush barbed wire, and support infantry attacks. Over time, other armies like the French and Germans also built their own versions, learning from each success and failure.

# Main Features of a Tank

1. **Tracks**
   Unlike cars, tanks run on caterpillar tracks that distribute their weight across a larger surface area. This helps them move over mud, sand, and uneven terrain without getting stuck as easily as wheeled vehicles would.
2. **Armor**
   Tanks are covered in thick metal or composite plating to protect against bullets, shells, and explosive charges. Over time, new materials have been used, including steel alloys, ceramic layers, and composite armor. The goal is to keep the crew safe while keeping the tank from being too heavy.
3. **Main Gun**
   Tanks usually have a large cannon that can fire shells capable of destroying other tanks, buildings, or fortified positions. Different shells can be loaded for different targets, including high-explosive or armor-piercing rounds.
4. **Machine Guns**
   In addition to the main cannon, tanks typically have smaller machine guns to defend against infantry. Some of these are mounted next to the main gun, while others may be on top of the turret.
5. **Turret**
   The turret is a rotating structure on top of the tank. Inside it, the gunner and commander can aim the main cannon in nearly any direction without moving the entire vehicle.
6. **Engine**
   Tanks need a powerful engine to drive their heavy frames. Diesel or gas turbine engines are common. These engines must be large enough to move the tank's weight at a decent speed while also powering systems like the turret rotation.

## Types of Armored Vehicles

Not all armored vehicles are tanks. Many serve different purposes:

- **Armored Personnel Carriers (APCs)**
  These are vehicles designed to move troops safely. They have armor to stop bullets and shrapnel, but they do not always have a big main gun. Instead, they may have a light machine gun or cannon for basic defense.
- **Infantry Fighting Vehicles (IFVs)**
  Similar to APCs, but they carry heavier armament, like an auto-cannon or anti-tank missiles. They let infantry fight from inside or alongside the vehicle.
- **Tank Destroyers**
  These vehicles specialize in taking out enemy tanks. They usually have a strong gun but might have less armor or a fixed turret, making them cheaper or easier to build.
- **Reconnaissance Vehicles**
  Fast and lightly armored, these vehicles scout enemy positions and gather intelligence rather than engage in heavy combat.
- **Self-Propelled Guns**
  Some tanks look similar to self-propelled artillery, but self-propelled guns are built more for firing powerful shells from a distance rather than direct tank-to-tank combat.

---

## How Tanks Fight

Tanks are often used in groups called armored units. They move together, supporting each other and the troops on foot. Modern tactics use the speed and power of tanks to break through enemy lines, while infantry clears out any threats like enemy soldiers with anti-tank weapons.

When tanks engage other tanks, they rely on aiming, armor, and maneuverability. The crew must find weak spots in the enemy's armor—often the sides or rear. Meanwhile, the team tries to keep the front of their tank facing the enemy, as that part is usually the strongest. Tank battles can be intense, as each side tries to outflank or surprise the other.

## Inside a Tank

A typical tank crew has four main roles:

1. **Commander**
   The commander is in charge of the tank. They decide where to go, when to fire, and how to coordinate with other units.
2. **Gunner**
   The gunner aims and fires the main cannon, using sights that may include magnification or thermal imaging for nighttime or bad weather.
3. **Loader**
   The loader puts shells into the main cannon. Some modern tanks use an automatic loading system, which replaces the need for a separate loader, but many still have this role.
4. **Driver**
   The driver steers the tank. They have their own station near the front. They often rely on the commander for direction, especially in places with poor visibility.

The space inside a tank is cramped. The crew has to work together in close quarters while dealing with heat, noise, and the rough movements of the vehicle. Modern tanks sometimes have air conditioning or advanced ventilation systems, but it can still be uncomfortable compared to an ordinary car.

# Armor Technology

Armor is crucial for keeping the crew safe. Early tanks just used thick metal plates. Modern tanks have advanced armor layers:

- **Composite Armor**
  Made of different materials like steel, ceramics, and special fibers. Each layer helps absorb and spread out the energy from an incoming round.
- **Reactive Armor**
  Blocks or tiles on the tank's exterior contain explosives. When a projectile hits, the explosive charge in the tile goes off, pushing the armor plate outward. This disrupts the force of shaped charges or high-explosive anti-tank (HEAT) rounds.
- **Active Protection Systems**
  Sensors detect incoming rockets or missiles. The system fires small projectiles or blasts to destroy or divert the threat before it hits the tank. This is a newer technology that can greatly increase survival chances.

These armor advancements let tanks survive hits that would destroy older models. Still, no tank is completely safe. A direct hit from a strong anti-tank missile or a high-powered cannon can knock out even advanced vehicles if it strikes a vulnerable area.

---

# Mobility and Engines

A tank's weight can exceed 40 to 70 tons or more, depending on its armor and size. To move this mass, it needs a powerful engine, often generating hundreds or thousands of horsepower. Tracks spread the tank's weight and give extra traction on mud, sand, and snow. However, tanks are not known for their high speed—many travel around 25 to 35 miles per hour on roads, though some modern tanks can go faster.

Fuel is a major concern. Tanks consume a lot of fuel, and a long mission can require frequent resupply. This makes logistical support very important. If a tank runs out of fuel or parts, it becomes a giant metal box stranded in the field.

## Tank Tactics Over Time

1. **World War I**
   Tanks were first used to cross trenches. Their surprise factor was huge, but mechanical failures limited their success.
2. **World War II**
   Tanks became central. Battles such as those on the Eastern Front between Germany and the Soviet Union showed how large tank formations could swing the tide of war. Countries improved tank designs, focusing on better guns, thicker armor, and more reliable engines.
3. **Cold War**
   Many nations continued to develop faster, stronger tanks, expecting major clashes in Europe. Tanks grew heavier and more complex, with better fire control systems.
4. **Modern Conflicts**
   Tanks remain important, but they face new challenges such as smart missiles and drones that can target them from the sky. Armies focus on combining tanks with other forces—infantry, air support, and artillery—to handle all threats.

## Armored Personnel Carriers and Infantry Fighting Vehicles

While tanks focus on heavy firepower, APCs and IFVs prioritize moving soldiers. They are usually lighter than main battle tanks, but they offer more seats and firing ports for troops. IFVs have stronger weapons than APCs, letting them engage lighter vehicles or dug-in enemies. The presence of APCs or IFVs means soldiers can move safely across dangerous areas, protected from small arms and shrapnel.

## Upkeep and Logistics

Armored vehicles need constant attention:

- **Maintenance**
  Tracks wear out, engines need regular checks, and all the mechanical systems must work smoothly. Turret rotation systems and gun stabilizers also require inspection and tuning.
- **Repairs**
  Battle damage, rough terrain, and heavy use can break parts. Many armies have special recovery vehicles built on tank chassis to tow damaged tanks to repair sites.
- **Upgrades**
  As technology improves, armies upgrade older tanks with better engines, electronics, or armor packages. This can extend a vehicle's service life by many years.
- **Fuel and Ammo**
  A single tank might carry dozens of shells for its main gun and thousands of rounds for its machine guns. Transporting all this equipment is a huge task that depends on reliable supply lines.

## Vulnerabilities and Threats

Tanks have strong armor, but they are not invincible. If an enemy has anti-tank missiles, mines, or attack aircraft, they can still destroy or disable a tank. Tanks also struggle in narrow urban streets, where high-rise buildings allow ambushes from above. Drivers must watch for improvised explosives (IEDs) hidden along roads in some conflict zones.

Because of these risks, armies often move tanks with infantry support. Foot soldiers protect the tanks from close-range threats, while tanks help the soldiers deal with bunkers or vehicles. This teamwork approach helps each side cover the other's weaknesses.

## Modern Armored Warfare

Today, tanks and armored vehicles are part of combined arms operations. This means they work with helicopters, jets, drones, artillery, and foot soldiers to fight as a coordinated team. Communication networks let a tank commander speak with pilots or artillery units in real time. The tank can share target coordinates, request support, or receive warnings about threats ahead.

Meanwhile, new tools like thermal sights and advanced optics help gunners find targets even at night or in bad weather. Computerized fire control systems can track moving objects and calculate aiming details automatically. Some tanks even have sensors that detect incoming fire, telling the crew where shots are coming from.

## Ethical and Economic Factors

Tanks and armored vehicles cost a lot of money. Building one can involve advanced factories, skilled workers, and specialized materials. This means not every nation can afford a large number of modern tanks. Some countries rely on older or imported vehicles, while wealthier nations invest in the newest designs.

There are also questions about using such force in populated areas. A tank's main gun can destroy buildings and harm civilians if it is fired carelessly. Armies are expected to follow rules to protect bystanders, especially in towns and cities.

## Future of Tanks and Armored Vehicles

While some people have suggested that tanks might become less important due to new weapons like drones or missiles, many armies still see a place for heavy armor on the battlefield. Tanks provide a mobile, well-protected base of firepower that smaller vehicles cannot match. The designs might change, though. Some future ideas include:

1. **Lighter Materials**
   Engineers continue to research metals and composites that provide the same protection at reduced weight.
2. **Hybrid or Electric Engines**
   Rising fuel costs and logistical concerns might push the development of alternative power sources.
3. **Unmanned Vehicles**
   Some prototypes remove the crew entirely, letting operators control the vehicle remotely. This could reduce the risk to human soldiers.
4. **Integrated Systems**
   Tanks will likely become even more linked to drones, satellites, and digital networks. Data sharing in real time will guide them to targets or warn them of threats.

Still, as technology changes, so do the threats. Enemies may design better missiles or weapons to disable even advanced armor. Thus, the race between firepower, armor, and tactics goes on.

# CHAPTER 7

## NAVAL SHIPS AND THEIR WEAPONS

Naval ships are large vessels used by navies around the world to protect coastlines, transport troops, and project power across oceans. They vary in size and function, from small patrol boats to enormous aircraft carriers. What makes these ships stand out are the weapons they carry, including guns, missiles, and torpedoes. In this chapter, we will look at the different types of naval ships, how their weapons work, and why they remain a mainstay of sea-based defense and offense.

## Early Uses of Warships

Navies have existed for a long time. In ancient times, civilizations like the Egyptians, Greeks, and Romans built war galleys—wooden ships powered by oars and sails. These vessels carried soldiers across the water and used simple weapons such as catapults or archers firing arrows. Later, as shipbuilding improved, nations added cannons to sailing ships. Battles at sea became a test of maneuvering and artillery skill.

During the Age of Sail, strong navies controlled trade routes and colonies around the world. Wooden ships with many cannons on each side fought in large formations. Commanders tried to position their ships so they could fire broadsides—where all the cannons on one side of the ship fired at the same time—inflicting massive damage on enemy vessels. This period showed how crucial a strong navy could be for protecting a nation's interests overseas.

The switch from wooden hulls to steel hulls, along with the move from sail power to steam engines, changed warship design dramatically. Ships became faster and sturdier, able to handle bigger guns. This helped lead to the modern battleships of the early 20th century, which were floating fortresses covered in thick armor.

## Types of Modern Naval Ships

Modern navies use many kinds of vessels, each designed for a specific job. Below are some of the most common types:

1. **Aircraft Carriers**
   These are among the largest warships in the world. An aircraft carrier has a full-length flight deck where fighter jets and helicopters can take off and land. The carrier itself might have only a few defensive weapons, relying on its air wing and escort ships for protection. Its main power comes from the planes it carries.
2. **Destroyers**
   Destroyers are fast and maneuverable ships meant to protect bigger vessels, like aircraft carriers, from threats such as submarines and incoming missiles. They carry surface-to-air missiles, anti-submarine weapons, and guns capable of bombarding targets on land.
3. **Cruisers**
   Cruisers are larger than destroyers and often carry a wide range of weapons, including cruise missiles. They can perform tasks like air defense, striking land targets, and hunting submarines. Some navies no longer use cruisers, but others keep them in service as key escorts for carriers.
4. **Frigates**
   Frigates are usually smaller than destroyers and often specialize in anti-submarine warfare. They might also escort convoys or protect merchant ships. Many navies depend on frigates for coastal defense and patrol duties.

5. **Corvettes**

    Corvettes are even smaller than frigates and are designed mostly for patrol and near-shore work. They can carry anti-ship missiles, torpedoes, and small guns. They are popular for navies that operate close to their own coastlines and need a flexible, fast craft.

6. **Amphibious Assault Ships**

    These large ships carry Marines or other forces for landing operations. They have well decks for launching smaller landing craft, and some can launch helicopters or vertical-takeoff aircraft. They often look similar to small aircraft carriers but are built for landing troops ashore.

7. **Submarines**

    Submarines are not surface ships, but they are still an essential part of a navy's power. They can move underwater to avoid detection, strike enemy ships with torpedoes, or launch cruise missiles at land targets. Some submarines carry nuclear missiles, acting as a deterrent.

8. **Patrol Boats**

    On the smaller end of the scale, patrol boats guard coasts and waterways. They may have machine guns, small cannons, or light missiles for intercepting smugglers, pirates, or any intruders.

Each of these ships plays a role in a navy's strategy, whether that is defending coastlines, controlling sea lanes, providing air support, or carrying out surprise attacks.

---

## Weapons on Naval Ships

Naval ships are platforms for a variety of weapons. Let us explore some major categories:

1. **Guns and Cannons**
   Modern warships often have at least one main gun, typically in the 76 mm to 155 mm range. These guns can fire shells at fast-moving sea targets or bombard enemy positions on land. Some vessels, like older battleships of the past, carried huge guns (up to 18 inches in caliber), but most modern designs focus on missiles rather than very large guns. Still, the naval gun remains important for close-range defense and shore support.
2. **Missiles**
   Guided missiles are a key weapon at sea today. They come in several forms:
   - **Surface-to-Air Missiles (SAMs)**: Protect the ship and its group from incoming aircraft or missiles.
   - **Surface-to-Surface Missiles (SSMs)**: Target enemy ships or land-based targets. These can travel long distances and often follow a path close to the water to avoid detection.
   - **Cruise Missiles**: Designed to fly low and far, sometimes hundreds of miles, to strike land targets with high accuracy.
   - **Ballistic Missiles**: Carried by some submarines or special cruisers. They can be nuclear-armed and launched in an arc that exits the atmosphere before re-entering to hit a distant target.
3. **Torpedoes**
   Torpedoes are self-propelled underwater weapons that explode against or near enemy ships or submarines. They are commonly fired by submarines and surface ships. Modern torpedoes can home in on their targets using sonar. Some can even be dropped from helicopters or planes.
4. **Close-In Weapon Systems (CIWS)**
   Many ships have CIWS to shoot down incoming missiles or aircraft at close range. A common example is a rapid-fire gun

that detects and tracks targets automatically with radar, firing many bullets per second to form a "wall of lead." This last line of defense can mean the difference between survival and destruction in a missile attack.

5. **Anti-Submarine Weapons**

    Besides torpedoes, ships may carry depth charges or specialized rocket-propelled charges that detonate underwater to damage or destroy enemy submarines. Modern systems often rely on helicopters or drones that drop sonobuoys (floating sensors) to find submarines, then direct torpedoes toward them.

6. **Electronic Warfare and Decoys**

    Naval ships also have electronic weapons to jam enemy radars or communications. They launch decoys to mislead incoming missiles, including chaff (tiny metal strips) and flares that confuse radar or heat-seeking devices.

---

## The Role of Aircraft Carriers

Aircraft carriers deserve special attention because they change how naval battles are fought. A carrier's main strength is its air wing: the fighter jets, bombers, and helicopters it carries. These aircraft can strike targets far away, protect the fleet from enemy planes, or help hunt submarines. Carriers often sail with a group of escort ships—destroyers, cruisers, and sometimes submarines—to shield them from threats.

Operating a carrier requires large crews, specialized training, and a clear command structure. Planes must take off and land on the ship's deck, which can be very difficult when the carrier is moving or weather is rough. Crews must handle refueling, arming, and maintaining the aircraft in a limited space. Carriers might also have advanced radar and communications systems to direct air missions.

Because they project power far from a nation's shores, aircraft carriers are often seen as symbols of naval strength. Not all countries can afford carriers, but those that can usually see them as a core part of their strategy. Critics note that carriers are expensive and might be vulnerable to modern missiles, especially if they sail too close to hostile coasts. Still, many naval planners argue that carriers are an unmatched way to control the skies over the ocean.

## Submarines as Part of the Fleet

Submarines play a dual role in naval warfare: stealth offense and nuclear deterrence. Conventional submarines (diesel-electric) run quieter at slower speeds, using batteries underwater. Nuclear-powered submarines can stay submerged for long periods, limited mostly by the crew's supply of food. By hiding underwater, a submarine can sneak up on enemy ships and launch torpedoes or missiles.

Some submarines carry nuclear-armed ballistic missiles. These "boomers" are strategic deterrents. By hiding in the depths of the ocean, they ensure that if an enemy attacked their home country, they could still strike back. This possibility of retaliation helps deter major conflicts. Submarines can also gather intelligence, drop special forces, or perform surveillance near enemy coastlines.

## Command and Control at Sea

A navy is a complex organization that must coordinate many ships, submarines, and aircraft across vast distances. Communication systems on board naval vessels are vital for passing orders, receiving intelligence, and sharing target data. Modern fleets use satellite links, radar networks, and secure digital channels. Automated systems can help track multiple targets at once, especially in crowded environments where many friendly and enemy units may be present.

Naval commanders must plan operations based on factors like geography, weather, and the specific capabilities of each vessel. In some regions, shallow or narrow waters make it easier for small boats or submarines to stage ambushes. In open oceans, storms and rough seas can challenge even large warships. Successful commanders combine technology, tactics, and training to overcome these obstacles.

## Life on a Naval Ship

Life on a naval ship can be demanding. Crewmembers live in tight quarters, often sharing small bunk areas. On a big ship like an aircraft carrier, thousands of people work in shifts, keeping the vessel running around the clock. Duties range from operating the reactors (if it is nuclear-powered) to cooking meals for the crew. Storms can make the ship roll, causing sea sickness for those not used to it.

Safety drills happen frequently. Fires and flooding are serious dangers at sea. Each person on board has a specific role in damage control, whether that is sealing off compartments or fighting flames. Maintenance is constant. Machinery must be inspected, cleaned, and repaired to prevent breakdowns.

Despite these challenges, many crewmembers find pride in their work, seeing themselves as part of an important mission. Naval service can involve long deployments away from home, but it also forges strong bonds among shipmates who depend on each other for safety.

# Naval Strategy and Power Projection

Naval power projection is about sending a strong signal to other countries. A navy might patrol international waters or conduct exercises near an area of tension to show presence and discourage aggression. Warships can also enforce blockades, stopping cargo ships from entering or leaving a certain region.

In large-scale conflicts, controlling the seas can be crucial. Nations need to secure shipping routes that carry goods like food and fuel. A strong navy helps keep these routes open and prevents the enemy from cutting off supplies. Sea control can also enable amphibious assaults, where troops land in hostile territory, supported by naval firepower.

On the flip side, a country with a weaker navy must develop strategies to protect its coastline using mines, missile boats, or land-based aircraft. Coastal defense can be strong if used wisely, forcing larger navies to be cautious near shore.

---

## Modern Advancements

1. **Stealth Technology**
   Some new warships use special angles and coatings to reduce radar visibility. By minimizing the reflection of radar signals, these ships are harder to detect. This approach is similar to what we see in stealth aircraft.
2. **Automated Systems**
   Computers handle many tasks, from controlling the engines to managing weapons. This can reduce the number of crew needed. However, it also means navies rely more on software, which could be hacked or malfunction.
3. **Unmanned Vessels**
   Experiments with unmanned surface and underwater

vehicles could change naval warfare. These drones might perform tasks like scouting or hunting submarines without risking a human crew.

4. **Integrated Data Sharing**
Ships, submarines, aircraft, and satellites share sensor data in real time. This networked approach helps a fleet identify and track targets much faster, allowing coordinated strikes.

5. **New Propulsion Methods**
Beyond nuclear power, some navies look at electric or hybrid propulsion for quieter and more fuel-efficient operations. This can extend a ship's range and reduce the amount of fuel it must carry.

---

## Balancing Cost and Capability

Building and running a modern navy is expensive. Aircraft carriers cost billions of dollars, and even smaller ships like frigates come with high price tags. The weapons, electronics, and skilled crews all add to the cost. Countries must weigh these expenses against other needs, like social programs or land-based defense.

As a result, some navies focus on smaller, multipurpose ships that can handle a variety of tasks. Others, like major world powers, invest in large fleets with carriers, cruisers, and submarines to maintain a global presence. Alliances between nations can spread the burden. For instance, some countries coordinate missions with allies who have complementary strengths, sharing resources and responsibilities.

---

## Environmental and Legal Considerations

Warships travel across international waters, which are regulated by treaties and rules. Navies must respect territorial boundaries and rights of passage. They also must consider environmental impacts, as large ships can carry oil and other materials that could spill in an accident. Sonar use, for example, can harm marine mammals that rely on sound for navigation.

International law tries to reduce tensions at sea. Treaties cover everything from how submarines must surface in narrow straits to rules about warning shots. Still, disputes happen when countries disagree over maritime boundaries or resource rights. A naval presence can calm or inflame tensions, depending on how it is used.

## The Future of Naval Warfare

Naval forces will likely keep adapting to new threats and technologies. Hypersonic missiles that travel many times the speed of sound pose a major challenge to existing defenses. Drones and robotic submarines might play a bigger role, adding to the complexity of detecting and countering underwater threats. Some believe lasers and directed energy weapons could appear on ships, providing a cost-effective way to stop incoming missiles.

Also, the need to protect trade routes remains vital. As global shipping continues, navies may place more focus on anti-piracy operations or safeguarding choke points like canals and straits. Climate changes, such as the opening of Arctic sea routes, could shift where navies operate.

Even with these new factors, the basic idea of a navy—to exert power at sea and defend a country's maritime interests—will remain. Ships will continue to carry out patrols, respond to disasters, support peacekeeping missions, and, if needed, fight wars on the high seas.

# CHAPTER 8

## AIRCRAFT AND THEIR WEAPON SYSTEMS

Aircraft have revolutionized warfare by bringing speed, height, and range to the battlefield. From early biplanes that dropped small bombs by hand to modern jets capable of precision strikes, airplanes and helicopters have become a vital part of armed forces worldwide. In this chapter, we will look at the main types of military aircraft, the weapons they carry, and how they fit into overall defense plans.

## The Birth of Military Aviation

The idea of fighting from the air began with observations balloons in the late 1700s. Armies used them to see the battlefield from above. Early aircraft in the early 1900s were not very reliable, but that changed rapidly with World War I. Pilots first used planes to spot enemy troops or direct artillery fire. Soon, they added small guns and bombs, turning planes into weapons themselves.

Aircraft quickly proved their worth, and countries raced to build better engines, stronger airframes, and more effective armaments. By World War II, bomber planes could fly long distances to strike factories or cities, and fighter planes tried to stop them. This air duel became a central part of modern warfare, with large-scale bombing raids shaping the conflict's outcome.

## Types of Military Aircraft

1. **Fighters**
   Fighters are fast planes built for air-to-air combat. They carry missiles and cannons to shoot down enemy aircraft. Modern fighters also attack ground targets with bombs or guided

weapons. Stealth fighters, like some used by major world powers, use special shapes and coatings to reduce radar detection.

2. **Bombers**

   Bombers are designed to carry heavy bomb loads over long distances. Strategic bombers can fly between continents, dropping bombs (possibly nuclear) on enemy targets. Tactical bombers, or strike aircraft, focus on battlefield or near-frontline targets.

3. **Attack Aircraft**

   Attack aircraft support ground forces by hitting enemy tanks, artillery, or positions. They fly lower and slower than fighters, aiming for precise hits. Some are heavily armored to survive anti-aircraft fire.

4. **Transport and Cargo Planes**

   These aircraft move troops, vehicles, and supplies. Large transport planes can land on rough airfields or drop cargo by parachute. They are not heavily armed but are crucial for logistics and humanitarian missions.

5. **Helicopters**

   Helicopters can hover, land in tight spots, and take off vertically. Attack helicopters carry cannons, rockets, or anti-tank missiles. Transport helicopters move soldiers quickly to remote areas. Some also conduct search and rescue or medical evacuation.

6. **Maritime Patrol Aircraft**

   These planes fly over oceans, tracking submarines or ships. They can drop sonobuoys to detect underwater activity or launch torpedoes. They are essential for navies that must guard large swaths of sea.

7. **Reconnaissance and Surveillance Aircraft**

   Using cameras, radar, or other sensors, these aircraft gather intelligence. High-flying spy planes can map enemy areas from great altitudes. Drones also fit into this category, though they have no pilot on board.

8. **Aerial Refueling Tankers**
   These planes carry extra fuel and refuel other aircraft in flight, extending their range. This lets fighters or bombers stay airborne longer and reach targets far away.

---

## Aircraft Weapons and Systems

Military aircraft rely on a range of weapons:

1. **Machine Guns and Cannons**
   Early planes had simple machine guns timed to fire through the propeller. Modern fighters may have an internal cannon for close combat if missiles fail. Attack helicopters also use cannons, sometimes in a swiveling turret under the nose.

2. **Air-to-Air Missiles**
   Fighters use short-range missiles guided by heat-seeking sensors for dogfights. For longer ranges, they use radar-guided missiles. These can travel many miles, locking onto targets using radar signals.

3. **Air-to-Ground Missiles**
   Designed to hit tanks, buildings, or other ground targets. Some have laser-guidance, meaning a laser designator is pointed at the target and the missile follows that beam. Others might use GPS or television guidance.

4. **Bombs**
   Bombs can be "dumb," meaning they drop straight, or "smart," guided by laser, GPS, or other signals. Smart bombs are more accurate, reducing collateral damage. Aircraft can carry dozens of bombs, depending on the plane's size.

5. **Rockets**
   Rockets are unguided projectiles fired in salvos to saturate an area. Attack helicopters and some fighter planes use them for close air support. While less precise than guided missiles,

they can be effective against groups of enemy vehicles or fortifications.

6. **Torpedoes and Depth Charges**
   Maritime patrol planes and helicopters may drop torpedoes or depth charges to attack submarines. After being dropped, a torpedo continues underwater, guided by sonar to find its target.
7. **Electronic Warfare Pods**
   Some aircraft carry pods that jam enemy radar or communications. Others may drop chaff and flares to mislead incoming missiles. This form of combat does not destroy targets physically, but it can disrupt or blind an enemy force.

## How Aircraft Fight in the Sky

Air combat can happen at close range, called dogfighting, or at long range, where pilots fire missiles guided by radar data. Fighters use tactics like turning maneuvers to get behind an opponent. Modern jets can detect targets far away with advanced radar, seeking to shoot before the enemy even knows they are there. Yet stealth technology aims to reduce radar detection, making it harder for an opponent's systems to lock on.

In ground-attack missions, pilots often rely on intelligence about enemy positions. They may fly low to avoid radar or remain high if the enemy has strong anti-aircraft defenses. Attack helicopters work closely with ground troops, using advanced sensors to spot threats hidden by terrain. Timing is crucial; aircraft might strike a position just before friendly forces move in.

Coordination with ground units or naval forces is also key. Aircraft might get target data from a forward observer, satellite images, or signals from a naval ship. Once the strike is complete, they can send back confirmation or video of the damage.

## The Importance of Air Superiority

Air superiority means controlling the sky so that an enemy cannot safely fly. This can change the outcome of a conflict. If one side's air force is stronger, they can protect their ground forces and bomb enemy troops with fewer worries. The weaker side either has to hide, use anti-air weapons, or risk severe losses.

This has led many militaries to invest heavily in fighter jets, surface-to-air missile systems, and robust training. Losing the air battle often means the opponent can strike ground targets freely, leaving defenses exposed. Even if a country cannot match an enemy's air force plane-for-plane, it might try to create strong anti-air defenses with missiles and radar to deter attacks.

## Helicopters in Modern Warfare

Helicopters are unique because of their vertical takeoff and landing ability. They can hover and land in spots airplanes cannot. Attack helicopters, like some well-known models around the world, can destroy tanks using anti-armor missiles. They hide behind hills and pop up to fire, using the terrain for cover. Transport helicopters insert troops into inaccessible areas, supply them, and evacuate the wounded.

However, helicopters are slower than jets and can be vulnerable to ground fire if they fly too close to enemy forces. Pilots may use nap-of-the-earth flying, staying just above treetops or hills to avoid radar. They also rely on flares and maneuvers to dodge shoulder-fired missiles.

## Role of Drones and Unmanned Aircraft

Unmanned aerial vehicles (UAVs), commonly called drones, are aircraft without a pilot on board. They come in many sizes, from small quadcopters used by ground units to large, long-endurance drones that can stay aloft for hours or days. Some are armed with missiles or bombs, while others focus on surveillance.

Drones are often used for missions considered too dull, dirty, or dangerous for human pilots. They can spy on enemy positions without risking a pilot's life. Armed drones can strike targets quickly once approved by a controller stationed far away. This can be very effective against groups hiding in remote areas.

Critics raise concerns about errors or accidents. A drone operator might misidentify a target if intelligence is flawed. There are also legal and ethical debates around using armed drones in countries that have not declared open conflict. Nonetheless, drones have become a common tool for many militaries.

## Aircraft Carriers and Naval Aviation

In the previous chapter, we discussed aircraft carriers, but let us review their aviation aspect. Carrier-based aircraft require special training for takeoffs and landings on a moving ship. Such planes often have stronger landing gear and tail hooks that catch cables on the deck to stop quickly. Naval fighters can protect the fleet from enemy aircraft, while strike aircraft can attack ships or land targets.

Helicopters also operate from smaller ships like destroyers or frigates. They help with anti-submarine warfare, scouting, and rescuing sailors who fall overboard. This flexibility allows a navy to extend its reach far beyond the coastline.

## Air Force, Navy, and Army Coordination

Modern warfare demands close coordination among air, land, and sea forces. An air force might handle strategic bombing or air-to-air combat, while the army's aviation branch (if it has one) provides direct support for ground units. The navy's aviation wing focuses on maritime missions, such as defending a carrier group or attacking enemy ships.

A single operation could involve satellites providing real-time images, drones scouting, and fighter jets launching precision strikes. Ground forces then follow up, and naval assets patrol offshore. Communication networks tie all these elements together, but if those networks fail or an enemy jams them, confusion can arise.

## Maintenance and Logistics for Aircraft

Aircraft need constant care to remain safe and mission-ready:

1. **Inspections**
   Pilots and mechanics check the plane or helicopter before and after every flight, looking for damage or wear. Engine parts, electronics, and control surfaces must be in good shape.
2. **Servicing**
   Planes need refueling and rearming. Larger air bases have facilities to load bombs, missiles, and machine gun ammo. For helicopters, ground crews handle rocket pods or turret cannons.
3. **Repairs**
   If a plane takes damage from enemy fire or experiences mechanical failure, technicians fix or replace parts. An engine overhaul may be done at a major depot. More routine maintenance might happen in the field.
4. **Upgrades**
   Over time, aircraft may receive better avionics, new sensors, or stronger materials. Software updates can also improve radar or targeting systems. This helps older planes keep up with modern threats.

5. **Supply Lines**
   Spare parts, fuel, and ammunition must flow smoothly to air bases. If supply lines are cut, aircraft might be grounded, limiting a military's air power.

## Electronic Warfare and Countermeasures

In modern air combat, technology can decide who wins. Fighter jets use radar to find enemy planes, but stealth aircraft try to reduce radar signatures. Electronic warfare involves jamming or tricking these radars. Pilots may drop chaff or flares to confuse incoming missiles. Advanced jets can also have towed decoys—small devices pulled behind the plane that attract enemy radar or missiles.

Pilots must constantly train to handle electronic warfare scenarios. They might fly with limited or no radar in "silent" mode to avoid detection, relying on allied aircraft or ground controllers for guidance. Battles can become a chess match of moves and counter-moves, with each side trying to outsmart the other's sensors.

## Strategic Bombing and Precision Strikes

Strategic bombing targets an enemy's ability to make war by hitting factories, refineries, or power plants. Nations may use heavy bombers carrying large payloads for these missions. Precision strikes, on the other hand, aim at specific spots like command centers or missile launchers. These strikes reduce unintended harm by using guided weapons that can pinpoint a target.

The debate around bombing centers on how it affects civilians. Even with guided bombs, mistakes happen. Some countries follow strict rules to minimize collateral damage, while others may be less careful. The rise of laser and GPS-guided munitions has made air attacks more accurate than in the past, but no system is flawless.

## The Future of Military Aviation

Aircraft technology keeps changing. Hypersonic missiles that travel many times the speed of sound challenge current defenses. Some prototypes explore directed-energy weapons, like lasers, that could shoot down incoming threats at close range. Next-generation fighters might be unmanned or have optional pilot control.

Stealth technology will likely advance, and new materials or shapes might hide aircraft even more effectively from radar. Drones will continue to grow in importance, possibly working in swarms to overwhelm defenses. Artificial intelligence could play a bigger role, helping planes make decisions quickly or fly autonomously in complicated situations.

Yet, the basics remain: aircraft need safe runways or landing areas, skilled crews, and good intel to hit the right targets. Weather and mechanical issues can still ground even the most advanced planes. A balanced approach—combining technology with sound tactics and well-trained personnel—often leads to success in the air.

## Balancing Air Power with Other Forces

A strong air force can deter aggression, protect troops, and strike enemies from afar. But planes alone rarely win a conflict. They need ground and naval forces to hold territory or control sea lanes. Aircraft carriers extend air coverage over oceans, while land-based jets can respond quickly to threats nearby. Helicopters fill in the gaps, working with soldiers on the ground.

In many cases, air power's biggest advantage is speed: jets can cross a region in minutes, offering a rapid response to emerging dangers. However, advanced anti-aircraft systems mean pilots face real risks, especially if they are forced to fly low. Balancing these factors is key to using aircraft effectively in any war plan.

# CHAPTER 9

## MISSILES AND ROCKETS

Missiles and rockets are weapons that move through the air (or even space) to strike targets from a distance. They can travel very quickly and often carry powerful warheads that explode on impact or near the target. While both rockets and missiles rely on thrust to push them forward, there is a key difference: missiles are guided, while rockets usually are not. In this chapter, we will look at how these weapons work, the many types that exist, and the effects they have on modern military strategy.

## What Are Rockets?

A rocket is a device that burns fuel to create hot gas, which then rushes out of the back at high speed. This push of gas propels the rocket forward. Rockets do not need air to function; they carry their own oxygen for burning fuel. This is why rockets can work in space, unlike typical airplane engines.

Rockets have many uses, from launching astronauts into orbit to delivering satellites. In a military setting, a "rocket" usually means a projectile without a guidance system. The path it takes depends on how it is launched, the angle, and the speed of its engines. Once fired, it travels on a predictable arc. Common military rockets include:

- **Artillery Rockets**: Fired in groups to hit enemy positions. These can saturate an area quickly, forcing enemies to take cover.
- **Rocket-Propelled Grenades (RPGs)**: Portable weapons used against vehicles or structures. They are small enough for one soldier to carry and fire.

- **Air-to-Ground Rockets**: Attached to aircraft or helicopters. Pilots aim the aircraft at the target and launch a burst of rockets.

Because rockets often lack sophisticated guidance, they can be less accurate. However, they are usually cheaper and easier to produce than guided missiles. This makes them valuable for armies that want to deliver a lot of firepower in a short time without spending too much money.

## Basic Parts of a Rocket

Although rockets come in many sizes, most share common parts:

1. **Propellant**: This is the fuel and the oxidizer that burn together to create thrust. Some rockets use solid propellant (a solid block of fuel), while others use liquid propellant stored in tanks.
2. **Motor Casing**: A strong shell that holds the fuel and provides structure.
3. **Nozzle**: The opening at the rear where hot gas escapes, pushing the rocket forward.
4. **Warhead**: The part that explodes upon impact (in a military rocket) or carries the payload in non-military rockets.
5. **Fins**: Small surfaces that help stabilize the rocket as it flies, making its path more predictable.

Some artillery rockets are placed in tubes on launch vehicles. A soldier or crew aims the vehicle and launches several rockets one after another, creating a barrage effect. The rockets themselves usually have a set flight path and will land in roughly the same area if launched in the same way.

# What Are Missiles?

A missile is a guided weapon. That means it has a system to adjust its flight while traveling. This system can include sensors, electronics, and moving surfaces that steer it toward the target. Missiles can be launched from the ground, ships, submarines, aircraft, or even from other missiles. They vary greatly in size and purpose, but they share the key idea of guidance.

Missiles can follow many paths, from a straight line close to the ground to a high arc through the upper atmosphere. Some missiles, called cruise missiles, fly more like airplanes, using wings and a jet engine to travel long distances at lower altitudes. Others, such as ballistic missiles, launch upward into space, then come back down at high speed toward their target.

---

# Guidance Systems in Missiles

Guidance is what sets missiles apart from simple rockets. Several types of guidance exist:

1. **Inertial Guidance**
   The missile has internal sensors (gyroscopes and accelerometers) that track its position and movement without relying on outside signals. It follows a pre-programmed path, making it hard to jam, but any small error in measurement can grow over long distances.
2. **GPS Guidance**
   The missile receives signals from satellites to determine its location. It adjusts its flight path to remain on course. GPS guidance can be very accurate, but if the enemy jams or disrupts GPS signals, the missile might lose its way.
3. **Laser Guidance**
   A laser designator on the ground or in an aircraft points a

laser beam at the target. The missile sees the reflected beam and steers toward it. This method can hit small targets accurately, but someone must keep the laser aimed correctly until impact.

4. **Infrared Guidance (Heat-Seeking)**

   The missile's sensor looks for the heat given off by engines or other warm parts. Many air-to-air missiles use this technology to home in on the hot exhaust of a jet. Flares or cool air can sometimes trick these sensors.

5. **Radar Guidance**

   The missile follows radar signals. In "active radar" missiles, the weapon carries its own radar transmitter and receiver. In "semi-active radar" missiles, the launch platform (like a plane or ship) illuminates the target with radar, and the missile locks onto the reflected signals.

6. **Command Guidance**

   The missile is steered by commands from an operator using radio or another data link. This can be very precise if the operator sees the target clearly, but it depends on keeping a communication link throughout the flight.

Each type of guidance has pros and cons. Some are better for short-range engagements, while others excel at hitting faraway or moving targets. The cost and complexity of guidance systems also vary. But in general, guided missiles allow a military to strike targets with greater accuracy and from safer distances than unguided rockets.

---

## Types of Missiles

Missiles are often named by their launch platform and target type. For instance, an **air-to-air** missile is fired from an aircraft to destroy another aircraft. A **surface-to-surface** missile is launched from the ground or a vehicle to hit ground targets. Let us explore some major groups:

1. **Surface-to-Air Missiles (SAMs)**
   Fired from the ground or ships to shoot down aircraft or incoming missiles. SAMs come in handheld forms (like a shoulder-fired weapon) or large systems with radar and multiple launchers. They can protect cities, bases, or ships from air attacks.
2. **Air-to-Air Missiles**
   Carried by fighter jets to defeat enemy planes in dogfights or at long ranges. Some have infrared seekers, while others rely on radar. Modern air forces rely heavily on these to gain control of the sky.
3. **Air-to-Ground Missiles**
   Launched from planes or helicopters to destroy tanks, buildings, or fortifications. Many use laser guidance or GPS. This category includes anti-tank missiles carried by attack helicopters.
4. **Anti-Ship Missiles**
   Designed to track and hit ships at sea. They can be fired from planes, ships, or shore-based launchers. Often they fly near the water's surface to avoid radar and can carry powerful warheads to pierce a ship's hull.
5. **Anti-Submarine Missiles**
   Some navies have missiles that can be launched from a ship or helicopter. They travel above water and then drop a torpedo into the sea near a suspected submarine.
6. **Surface-to-Surface Missiles**
   This group covers everything from short-range battlefield rockets to long-range ballistic missiles. They can be used to strike nearby enemy forces or distant targets like strategic bases.

7. **Ballistic Missiles**
   A special type of surface-to-surface missile that travels in a high arc. After launch, they coast through space before dropping back down. Some ballistic missiles can carry nuclear warheads over continents, making them key strategic weapons.

## Ballistic Missiles and Their Phases

Ballistic missiles are worth a closer look because of their range and destructive potential. A typical ballistic missile flight has three phases:

1. **Boost Phase**
   The missile's engines burn to escape the ground. If it is an intercontinental ballistic missile (ICBM), it might head toward outer space. In this phase, the missile is most visible to enemy sensors because of the hot exhaust.
2. **Midcourse Phase**
   The missile is in space or high in the atmosphere, coasting without the engine firing. Some missiles release a warhead or multiple warheads at this point. Decoys can also be released to confuse defenses.
3. **Terminal Phase**
   The warheads fall back into the atmosphere at high speed, heading toward the target. This is when defensive systems, like anti-ballistic missile networks, try to intercept them.

Ballistic missiles can travel huge distances, sometimes more than 5,000 miles. This long range often involves nuclear payloads, which we will discuss in the next chapter. Defending against them is hard. Anti-ballistic missile systems must detect, track, and destroy incoming warheads in a very short window of time.

# Rocket and Missile Launch Platforms

Weapons like rockets and missiles can be launched from various platforms:

1. **Ground Vehicles**
   Trucks with launchers can move quickly, fire, and hide. This mobility makes them harder to locate. Some smaller rocket systems are carried by soldiers on foot.
2. **Ships and Submarines**
   Many navies arm their vessels with missiles for defense or attack. Submarines can launch missiles while underwater, surprising the enemy. Warships may have vertical launch tubes that hold different missile types.
3. **Aircraft**
   Fighter jets, bombers, and helicopters carry missiles under their wings or in internal bays. This allows them to strike distant targets or enemy planes. Once launched, the aircraft can turn away while the missile continues on course.
4. **Fixed Silos**
   Some ballistic missiles sit in underground silos. These are heavily guarded and built to resist attack. However, they are not mobile, so they can be targeted if an enemy knows their location.
5. **Special Launch Pads**
   Some rockets, especially larger ballistic missiles, might be fired from a prepared site with fueling systems and strong infrastructure. This can be seen with space launch vehicles or older missile bases.

---

# Accuracy and Circular Error Probable (CEP)

When judging a rocket or missile, one measure is how close it lands to the target, often called **accuracy**. A common term is **Circular**

**Error Probable (CEP)**. This is the radius of a circle around the aim point where half the shots will land. For example, if a missile has a CEP of 10 meters, it means that half its warheads will land within a 10-meter circle around the target. A smaller CEP is better, meaning the weapon is more accurate.

Guided missiles tend to have smaller CEP values than unguided rockets. Some modern cruise missiles can hit within a few meters of their target from hundreds of miles away. This precision reduces the number of weapons needed and can lower unintended damage if used carefully.

---

## Defense Against Rockets and Missiles

As rockets and missiles became more common, militaries developed ways to stop them:

1. **Patriot and Similar Systems**
   Surface-to-air missile systems like the Patriot (used by some countries) intercept incoming ballistic or cruise missiles. They detect the threat with radar, then launch interceptor missiles.
2. **Naval CIWS**
   Warships rely on close-in weapon systems and surface-to-air missiles to shoot down incoming missiles. This is crucial in defending against anti-ship missiles.
3. **Iron Dome and Similar Shields**
   Some nations have built specialized systems to knock out short-range rockets fired at populated areas. Radar detects the rockets, and interceptor missiles destroy them before they land.
4. **Early Warning Radars**
   These giant radars can spot ballistic missiles soon after

launch. They give defensive systems extra time to prepare an interception, especially in the boost phase when the missile is easiest to track.
5. **Electronic Warfare**
   By jamming or confusing a missile's guidance signals, defenders may cause it to miss its target. This can involve interfering with radar, GPS, or data links.

While defenses can be effective against some threats, none are perfect. Large volleys of rockets or advanced ballistic missiles can overwhelm or outsmart defenses. This arms race continues as attackers improve their weapons and defenders refine interception methods.

## Tactical and Strategic Value

Rockets and missiles provide flexibility:

- **Tactical Uses**: On a battlefield, smaller rockets and short-range missiles can support ground troops, destroy enemy tanks, or target artillery positions. Unguided rockets can pound an area, while guided missiles pick off specific targets.
- **Strategic Uses**: Long-range missiles, especially ballistic ones, can threaten cities or bases far away. This can force the enemy to negotiate or deter them from attacking. Some missiles are meant for delivering nuclear warheads, making them a key part of a country's nuclear strategy.

The difference between tactical and strategic weapons is often the range and the type of warhead. Tactical weapons focus on a battlefield, while strategic weapons aim at larger political or military goals, sometimes crossing continents.

## Non-Military Rockets

Although this chapter focuses on military rockets, it is worth noting that rockets are also used for peaceful purposes, such as:

- **Space Exploration**: Launching satellites, sending probes to other planets, or putting astronauts into orbit.
- **Scientific Research**: Rockets carry instruments to high altitudes to study Earth's atmosphere or space phenomena.
- **Rescue Flares**: Small rocket devices can carry bright flares high into the sky for signaling help in emergencies.

Many breakthroughs in rocket technology came from space programs. Over time, these advances sometimes found their way into military designs, and vice versa. Still, the aims can be very different.

## Dangers and Ethical Questions

Rockets and missiles can cause a great deal of damage if used improperly. Their power to strike distant places can lead to civilian harm if they land in towns or cities. When militaries fire them in large numbers, the risk of mistakes or accidents grows, especially if guidance systems fail or if the targeting information is wrong.

There is also worry about these weapons ending up in the hands of groups that might use them against non-military targets. Smaller rockets are easier to smuggle, and some missiles can be sold on black markets if not controlled carefully. International treaties try to manage the spread of such weapons by limiting exports or requiring inspections. However, enforcing these rules is a constant challenge.

## Trends in Missile and Rocket Development

Developers continue to refine these weapons:

1. **Hypersonic Missiles**
   Traveling at speeds of Mach 5 or more, these missiles can make defense very hard. They can maneuver unpredictably, shortening reaction times for interceptors.
2. **Smarter Guidance**
   Artificial intelligence and improved sensors might give missiles the power to recognize and lock onto targets even if they move or change location.
3. **Smaller, More Mobile Launchers**
   Portable launch systems that can pop up, fire, and move away quickly are harder to track down. They give militaries a "shoot and scoot" option.
4. **Longer Range**
   Some missiles aim for extended reach, letting them strike deeper into enemy territory. This can shift military balances, as nations near the range of these new weapons feel threatened.
5. **Network-Centric Warfare**
   Missiles might network with drones, satellites, or each other. Sharing data in real-time can improve accuracy and enable coordinated strikes where multiple missiles attack from different angles.

# CHAPTER 10

## NUCLEAR WEAPONS AND THEIR EFFECTS

Nuclear weapons are among the most powerful and dangerous devices ever created. One explosion from a nuclear weapon can cause massive destruction, releasing deadly radiation and altering the environment for years. Only a few nations have these weapons, and much effort has been put into preventing their spread. In this chapter, we will discuss how nuclear weapons work, the results of their use in the past, and the serious risks they pose to people and the planet.

## What Makes a Nuclear Weapon Different

Conventional bombs rely on chemical explosives to create a blast. Nuclear weapons, by contrast, gain their explosive power from reactions that occur within atoms themselves. These reactions release far more energy than any chemical process can. As a result, nuclear weapons produce blasts many times stronger than typical bombs.

Two main types of nuclear reactions can power these weapons:

1. **Fission**
   Splitting atoms of heavy elements like uranium or plutonium into smaller parts. This process gives off a large amount of energy and additional neutrons that can split more atoms, causing a chain reaction.
2. **Fusion**
   Combining lighter elements like hydrogen into heavier ones (like helium). This releases even more energy than fission. Fusion is the same process that powers the sun.

Some nuclear weapons use only fission, called atomic bombs or A-bombs. Others combine fission and fusion, called hydrogen bombs or H-bombs, which can be far more powerful.

## Basic Parts of a Nuclear Warhead

Though nuclear weapons are complex, most share key components:

1. **Fissile Material**: Uranium-235 or plutonium-239 that can sustain a chain reaction. In a fusion-based warhead, there is also fusion fuel like deuterium or tritium.
2. **Explosive Lenses**: High explosives arranged around the fissile material. When they detonate, they compress the material into a critical mass, starting the chain reaction.
3. **Neutron Source**: A small device that emits neutrons at the right moment to trigger the chain reaction reliably.
4. **Tamper and Reflector**: Layers around the core that help keep neutrons from escaping, making the reaction more efficient.
5. **Casing**: The outer shell that holds everything together, sometimes shaped to increase the weapon's power (for instance, a reflective casing in a hydrogen bomb).

When set off, a nuclear warhead can release a huge burst of energy, in the form of blast, heat, and radiation. The exact amount depends on the design and yield of the weapon.

## The Effects of a Nuclear Explosion

A nuclear blast has several destructive effects:

1. **Immediate Blast Wave**
   Within seconds, a shock wave radiates outward, flattening buildings and throwing debris. In large explosions, this wave can demolish entire neighborhoods. People within the blast zone can be severely injured or killed instantly.

2. **Intense Heat and Fireball**
   The explosion creates a fireball that can reach temperatures hotter than the sun's surface. This heat can ignite fires, burn people's skin, and melt certain metals near the center of the blast.
3. **Radiation**
   Nuclear reactions release dangerous radiation in two main forms:
   - **Prompt Radiation**: High-energy rays (gamma rays and neutrons) emitted during the explosion.
   - **Fallout**: Radioactive particles that rise with the mushroom cloud and fall back to earth, potentially far from the explosion site, depending on wind patterns.
4. **Thermal Pulse**
   A strong flash of light and heat can cause burns or blindness. People and objects far from the blast center might experience severe burns if not protected.
5. **Electromagnetic Pulse (EMP)**
   A burst of electromagnetic energy that can damage electrical grids, electronic devices, and communication networks over a wide area.

A single nuclear weapon can kill thousands or even hundreds of thousands of people in a densely populated city. The damage to infrastructure—roads, hospitals, power lines—can be overwhelming, making rescue and medical care difficult.

## Historical Uses: Hiroshima and Nagasaki

The only time nuclear weapons were used in war was near the end of World War II. In August 1945, the United States dropped two atomic bombs on Japan—one on Hiroshima and one on Nagasaki. These bombs killed tens of thousands of people immediately, with many more dying later from burns and radiation sickness.

The shock of these events helped lead to Japan's surrender and ended World War II. However, the horror of the destruction began global debates on whether such weapons should be allowed to exist. Many feared that if more bombs were used, entire cities could be destroyed in a matter of moments.

## The Nuclear Arms Race

After World War II, several nations raced to develop and stockpile nuclear weapons. The United States and the Soviet Union led this drive, building thousands of warheads. Other countries, including the United Kingdom, France, and China, followed. India and Pakistan tested nuclear weapons, as well. Israel is widely believed to have them, although it has not confirmed this publicly. North Korea has conducted nuclear tests in recent years.

This competition led to the creation of bigger and more advanced warheads. Weapons were placed on missiles, submarines, and bombers, creating a constant threat that if one superpower attacked, the other would strike back. This idea, called "Mutual Assured Destruction," argued that neither side would start a nuclear war because both would be destroyed.

## Delivery Systems

Nuclear weapons can be delivered by many methods:

1. **Intercontinental Ballistic Missiles (ICBMs)**
   Launched from silos or mobile launchers, these missiles can travel thousands of miles, reaching other continents in about 30 minutes. They often carry multiple warheads that split off in space.

2. **Submarine-Launched Ballistic Missiles (SLBMs)**
   Fired from submarines hidden underwater. They are hard to detect and can appear near enemy shores, reducing warning times.
3. **Strategic Bombers**
   Heavy bombers, like certain aircraft used by major powers, drop nuclear bombs or fire cruise missiles. These planes can refuel in midair, extending their range.
4. **Cruise Missiles**
   Smaller and slower than ballistic missiles, but often able to fly low and avoid radar. Some nuclear-armed cruise missiles can be launched from ships, submarines, or planes.
5. **Tactical Weapons**
   Shorter-range bombs, shells, or missiles meant for battlefield use. Although "smaller," they can still be devastating.

Because these systems can strike with little warning, they create tension during conflicts. Each side wants to track the other's nuclear forces, so intelligence and surveillance become crucial.

---

## Radiation and Long-Term Health Effects

Radiation from a nuclear explosion can cause short and long-term harm:

- **Radiation Sickness**: High doses of radiation damage cells, leading to nausea, vomiting, hair loss, and internal bleeding. In severe cases, death can occur within days or weeks.
- **Cancer and Birth Defects**: Even lower levels of radiation can increase cancer risks or harm unborn babies. People exposed to fallout might develop health problems years later.
- **Environmental Damage**: Radioactive particles can contaminate soil and water. Plants and animals can also carry radiation in their bodies. This contamination can last for decades, depending on the materials involved.

Areas near bomb tests or accidents can remain dangerous for a long time. The region around Chernobyl, where a nuclear reactor accident occurred (not a bomb), still has high radiation levels decades later, showing the lasting impact of radioactive contamination.

## Attempts at Control and Disarmament

Realizing the terrible power of nuclear weapons, nations have tried to manage and reduce them:

1. **Nuclear Non-Proliferation Treaty (NPT)**
   An international agreement aimed at preventing the spread of nuclear weapons. Countries without them pledged not to obtain them, and countries with them promised to work toward disarmament.
2. **Comprehensive Nuclear-Test-Ban Treaty (CTBT)**
   Seeks to ban all nuclear explosions. Many countries signed, but it has not fully gone into effect because some key nations have not ratified it.
3. **Strategic Arms Reduction Treaties (START)**
   Agreements between the United States and Russia to cut down their nuclear stockpiles. These treaties limit the number of deployed warheads and delivery systems each side can have.
4. **Bilateral and Regional Treaties**
   Some countries form regional nuclear-weapon-free zones, banning nuclear devices in those areas. Examples include parts of Latin America and Africa.
5. **International Oversight**
   The International Atomic Energy Agency (IAEA) inspects nuclear facilities to ensure materials are used for peaceful purposes. It tries to detect any secret weapons programs.

Despite these efforts, disagreements remain. Some nations argue they need nuclear weapons for defense. Others say no nation should have them because of the destruction they can cause. The push and pull between these views continues to shape global politics.

## Modern Nuclear Strategies

Today's nuclear strategies often revolve around deterrence. Countries keep enough warheads to survive a first strike and retaliate, making it risky for anyone to launch a nuclear attack. This stand-off is believed by some to keep peace, while others see it as a dangerous gamble.

Nations also focus on **second-strike capability**—ensuring that even if their land-based missiles are destroyed, their submarines or bombers can still strike back. This is why many see submarine-based weapons as vital, since submarines can hide under the ocean.

Some countries also discuss lower-yield nuclear weapons that are less destructive, intended for limited strikes. Critics warn that this might make using them seem more acceptable, raising the risk of nuclear conflict.

## Potential Catastrophes

If a full-scale nuclear war broke out between major powers, the consequences would be grim:

1. **Immediate Casualties**
   Millions could die in the first hours from blasts and fires.
2. **Nuclear Winter**
   Soot and dust thrown into the atmosphere might block sunlight, cooling the planet for months or years. Crops could fail, causing widespread famine.

3. **Radiation and Disease**
   Radioactive fallout could spread around the globe, increasing cancer rates and poisoning water supplies.
4. **Loss of Civilization**
   Destruction of infrastructure—power grids, hospitals, transportation—could plunge survivors into chaos, with little ability to rebuild quickly.

Even a more limited nuclear exchange would cause major regional and global problems. The fear of this outcome has driven many leaders to avoid direct confrontations that could lead to nuclear use.

## Accidents and Near Misses

There have been accidents in the past that nearly led to disasters, often called **near misses**:

- **Accidental Launch**: A missile might launch due to a technical glitch or false alarm. During the Cold War, there were instances where radar systems mistook natural events for incoming missiles. Quick thinking by individuals sometimes stopped a mistaken counter-attack.
- **Broken Arrows**: This term refers to accidents involving nuclear weapons, such as when bombs are lost or damaged. Some bombs were dropped from planes by mistake, landing in oceans or near communities.
- **Unauthorized Use**: The risk that a rogue commander or a thief could grab a weapon and use it without permission. Nuclear devices usually have strict locks and security, but no system is foolproof.

Such incidents highlight how critical safety protocols are. Many layers of checks, codes, and secure communications exist to lower the chance of an accidental nuclear detonation.

## Civil Defense Measures

In the mid-1900s, some countries advised citizens to build shelters or practice drills for nuclear attacks. People were told to "duck and cover," hoping to reduce injuries from broken glass or flying debris. While these measures might help in smaller blasts, surviving a large-scale nuclear war remains challenging. Governments still have emergency plans, though they rarely discuss them openly, because the scenario is so dire.

Medical professionals also train to deal with radiation sickness, burns, and mass casualties. However, the scale of possible injuries could overwhelm any system. Many experts argue that the only true safety from nuclear weapons is not having them in the first place.

## Ongoing Concerns

While the Cold War ended decades ago, nuclear weapons remain a global concern:

1. **Rising Tensions**
   Some regions still face conflicts that could escalate, like border disputes between nuclear-armed nations.
2. **Non-State Actors**
   There is fear that terror groups could acquire radioactive materials for a "dirty bomb," though not as powerful as a true nuclear device, it could still spread contamination and panic.
3. **Modernization**
   Countries upgrade their arsenals with new missiles, warheads, and submarines, which can heighten competition.
4. **Arms Control Challenges**
   Treaties can break down if nations accuse each other of cheating. Also, new technologies (like hypersonic weapons) can complicate existing arms agreements.

As a result, world leaders meet in conferences to discuss ways to keep nuclear weapons in check, but progress can be slow.

## Ethical and Moral Debates

Nuclear weapons raise tough ethical questions. Some believe they keep the peace by deterring war, claiming past decades have not seen a world conflict on the scale of World War II partly because of nuclear fear. Others say it is immoral to threaten entire populations with mass destruction. Religious leaders, peace groups, and many public figures have called for a total ban on nuclear arms.

At the same time, nations with such weapons see them as insurance against invasion. So long as nuclear weapons exist, the risk of their use—by accident or design—lingers. The debate often asks if the risk is worth the perceived security. Many also point out the enormous costs of building and maintaining these arsenals, which might be spent on healthcare, education, or infrastructure instead.

## Possible Future Paths

The future of nuclear weapons could go in several directions:

1. **Continued Deterrence**
   The status quo remains, with nuclear-armed states keeping and modernizing their arsenals while non-nuclear states rely on other means of security or alliances.
2. **Gradual Disarmament**
   More treaties and agreements may reduce arsenals over time. This would require trust and verification among nations, plus public support.
3. **A New Arms Race**
   If global tensions rise, states might rush to build even more advanced nuclear weapons, ignoring treaties in favor of self-protection.

4. **Breakthrough in Defense**
   Perhaps technology advances to the point where ballistic missiles can be reliably intercepted, lowering the risk of nuclear attacks. However, that might push nations to find other ways to deliver warheads, like stealth submarines or hidden cruise missiles.
5. **Complete Ban**
   In a best-case scenario, all nations might agree to destroy their nuclear stockpiles, with strict inspections. Critics say this is unlikely as long as mistrust and power struggles remain.

Whatever happens, nuclear weapons will remain a topic of global importance. Their ability to cause incredible harm ensures that leaders and citizens alike keep a watchful eye on how they are managed.

# CHAPTER 11

## WHO MAKES WEAPONS AND WHY

Weapons come in many forms—from small arms carried by a single person to massive rockets that can cross continents. While governments often set the rules for how these items are used and traded, the actual work of making weapons is carried out by various groups of people and organizations. Some are huge companies that sign multi-billion-dollar contracts with governments. Others are smaller businesses with a handful of workers. In many cases, governments themselves also produce weapons in state-owned factories.

But who are these manufacturers, and why do they make weapons? In this chapter, we look at the groups involved in building weapons, the reasons they do it, and the people who take part in the process. We also explore how public opinion, government demands, and economic forces shape the decisions about what weapons get made and who buys them.

## The Role of Defense Companies

Defense companies—also called arms manufacturers—are businesses that design, produce, and sell weapons or equipment to militaries. Some of these companies focus on a single type of product, such as small firearms. Others produce a wide range of military goods, from fighter jets to communication systems. In many countries, the top defense companies are large and well-known, often among the biggest businesses on the stock market.

### Size and Scope

1. **Multinational Giants**
   Some defense companies operate in many countries, signing contracts with different governments. They can be involved

in projects for airplanes, missiles, ships, and even space systems. Their size allows them to handle huge contracts, but it also means they must follow many rules and deal with a lot of public attention.

2. **Specialized Medium Firms**

   These firms might focus on one product or service, such as drone technology or parts for a tank's engine. They may not be as large as the biggest companies, but they are important parts of the supply chain, making key components or offering unique expertise.

3. **Small Arms Manufacturers**

   Many smaller businesses produce rifles, pistols, and ammunition. Some of these cater to the police or civilian markets as well as the military. They may not have the huge budgets of the major players, but they still influence the industry.

**Government Contracts**

Most defense companies depend on government contracts for their income. A nation's military might request a new fighter jet or a naval ship, and the defense company will bid to build it. Because these deals can be worth billions of dollars, winning a contract is a big deal for a company. Governments usually pick a supplier based on technical ability, price, and track record of delivering on time. However, politics can also play a part. A government may prefer to award contracts to companies in its own country to protect local jobs and keep military secrets at home.

## Government-Owned Weapon Factories

In some countries, the government itself operates factories that build weapons, rather than relying on private companies. This approach can happen for a few reasons:

1. **Security**: By making weapons at state-owned factories, a government can reduce the chance that secret technology will leak.
2. **Control**: Relying on private firms means the government must sign contracts and share details. But if the state handles the production, it has tighter control over timelines, costs, and design.
3. **Strategic Needs**: During conflicts or emergencies, a government may feel it cannot rely solely on private firms that could be influenced by market conditions.

These state-owned operations still require skilled workers, materials, and technology. They often function similarly to private companies but without the need to make a profit. Instead, their success is judged by how well they fulfill the military's needs and stay within budgets set by the government.

## The People Behind the Scenes

Weapons are not built by machines alone. They come from the hands and minds of many people with different backgrounds:

1. **Engineers**: Responsible for designing everything from the shape of a rifle barrel to the guidance system of a missile. They use math, physics, and technology to solve problems, making sure a weapon performs reliably under stress.
2. **Technicians and Factory Workers**: Assemble parts, weld metal plates, fit gears, install computer boards, and run tests. These workers have hands-on roles, making sure each piece meets safety and quality standards.
3. **Scientists and Researchers**: Work in labs to study new materials, advanced electronics, or better propulsion methods. They might try to invent lighter armor, more accurate targeting devices, or powerful explosives.

4. **Managers and Executives**: Oversee budgets, schedules, and contracts. They balance the cost of parts and labor with the amount a client will pay. They also manage the relationships with government officials who decide which projects get funded.
5. **Sales and Marketing Teams**: Work on winning new contracts. They make proposals, attend trade shows, and handle negotiations with clients at home and abroad. Their success can be crucial for a company's survival.
6. **Legal and Compliance Experts**: Focus on rules about arms exports, patents, and safety regulations. Given the tight rules on weapons, these experts make sure the company does not break any laws.

## Why Do Companies Make Weapons?

At the most basic level, many businesses make weapons to earn money. Demand for military equipment can stay strong even in economic downturns, because countries often keep defense budgets high. But profit is not the only reason:

1. **National Security Roles**: Some manufacturers see themselves as part of a nation's defense framework. They feel a sense of duty to provide the best equipment for their country's armed forces.
2. **Innovation and Research**: Weapons research can drive new technology. A company may enjoy the challenge of working on advanced systems that eventually spin off into civilian products. For example, military spending on jet engines or communication satellites has led to progress in commercial aviation and space technologies.
3. **Influence and Prestige**: Building top-of-the-line fighter jets or submarines can boost a company's reputation. It may also

give the business a strong voice in political discussions, because governments rely on them for key projects.
4. **Employment**: Large defense contracts create many jobs. Politicians and public officials often value these contracts because they keep local economies strong, leading to a positive view of these companies in certain regions.

## Arms Dealers and Brokers

Beyond formal business deals, there is also an industry of brokers and dealers who help move weapons between countries. They arrange sales, handle paperwork, and sometimes connect buyers with sellers in places where direct contact is difficult due to distance or legal barriers. While many brokers obey laws, some may operate in the shadows, selling guns to banned groups or smuggling weapons into conflict zones. Stopping illegal arms dealing is a challenge for international agencies.

## Impact on Local Economies

Large factories that make tanks, airplanes, or missiles can become major employers in a town or city. This can shape local economies:

1. **Stable Jobs**: Defense work often pays well and can be seen as steady, because governments rarely stop buying weapons altogether.
2. **Skilled Workforce**: Businesses might invest in training or partner with schools to develop specialized courses in welding, machining, or electronics.
3. **Growth in Other Sectors**: As engineers and workers settle in the area, they spend money on housing, food, and services, boosting local businesses.

On the downside, these communities might depend too much on defense contracts. If the government cuts spending or picks a different supplier, factories can close, causing unemployment. Also, some people feel uneasy about living in a place tied so closely to building weapons.

## The Global Arms Industry

The arms industry is global. Some countries are leading exporters, while others import most of their military gear. Factors that shape these trades include:

1. **Alliances and Diplomacy**: Countries that share similar politics or are partners in defense treaties often buy weapons from each other.
2. **Price and Quality**: Nations on tight budgets might pick cheaper options, even if they are less advanced. Wealthier buyers may go for high-end technology.
3. **Offsets and Local Production**: A client might demand that part of the weapon be made in their own country, creating jobs and technology transfers. Some deals even include training local workers or setting up factories abroad.

The biggest exporters of major weapons—like fighter jets, tanks, and large naval vessels—are often the United States, Russia, certain European nations, and some East Asian countries. Buyers range from big militaries to smaller nations looking to upgrade their defense.

## Ethics and Public Opinion

Not everyone agrees that making weapons is a good line of work. Critics worry that mass-producing arms encourages conflicts or supports governments with poor human rights records. They ask if companies should be responsible when their products are used to harm civilians or stir up unrest.

On the other hand, supporters argue that strong defense industries help countries protect themselves, create jobs, and maintain a level of readiness. They point out that weapons are not always used for aggression; many nations rely on them to deter attacks or help peacekeeping missions.

Still, public opinion can shift. A scandal involving weapons sold to a regime known for oppression can hurt a company's image. This pressure might cause a manufacturer to rethink certain deals or set stricter guidelines for who can buy its products.

## Lobbying and Political Influence

Defense firms often spend time and money to influence politicians and government officials. They might hire lobbyists to explain why a new contract is good for jobs and security. In some countries, retired military officers or government leaders find jobs at defense companies, using their connections to help win contracts.

This closeness can raise questions about fairness. Some wonder if politicians approve deals because they benefit from them personally, rather than because the equipment is truly necessary or cost-effective. To address these concerns, laws in many places require transparency in government contracting. Yet, the arms industry remains known for its close ties to government power.

## Technology Sharing and Collaboration

Large weapons programs can be too costly for one country alone. This leads to collaborative projects, where multiple nations share expenses and the final product. A good example is a fighter jet program where different countries split the research tasks. Each partner nation then buys the jets and has a role in building components.

This approach saves money and can foster goodwill among the participating nations. However, it can also cause delays if partners disagree on design choices or budgets. Dividing up work across many countries can make production complex and slow.

## Privatization and Outsourcing

In past decades, many governments looked to privatize certain parts of their defense work. Activities that were once done by state-owned plants, like building trucks or basic gear, might be handed over to private firms. The idea is that private businesses may be more efficient or inventive. Critics argue that vital military needs should not be driven by profit motives, worrying that private firms might cut corners to save money.

At the same time, some militaries rely on private contractors for tasks like vehicle maintenance, logistics, or even armed security. This practice raises questions about accountability. When a private security company has employees in a conflict zone, do they follow the same rules as soldiers? Who is in charge if something goes wrong? These are ongoing debates in many countries.

## Arms as a Symbol of National Strength

A country with advanced weapons industries can be seen as more powerful on the world stage. Building sophisticated equipment—like stealth fighters or nuclear submarines—can boost a nation's status. This can encourage governments to pour money into arms projects, even when budgets are tight. They see it as a matter of pride or deterrence, showing potential adversaries that they can defend themselves.

However, focusing too heavily on arms production can divert resources from other needs, such as healthcare, education, or infrastructure. Leaders must decide how to allocate funds to balance security with the well-being of citizens.

## Black Markets and Unlicensed Production

Not all weapon makers follow legal paths. In some regions, small workshops or groups produce arms without official oversight. They might build copies of popular guns to sell on the black market. These unregulated weapons can end up in the hands of criminals or fighters in civil conflicts, making violence worse. Governments try to crack down on illegal arms factories, but it can be a never-ending chase. Local economies might support these operations because they provide jobs and income where legal businesses are scarce.

## Environmental and Safety Concerns

Weapon factories can pose environmental hazards if chemicals or metals leak into the ground or water. High-energy processes used in creating explosives or refining metals might produce toxic waste. Proper disposal and cleanup can be costly, and some companies have been accused of neglecting this duty. Workers also face risks on the job, especially in plants dealing with explosives. A single mistake can lead to an accident that injures or kills employees. Companies must follow strict safety rules, but accidents can still happen.

## Shifts in the Industry

The weapon industry changes over time:

1. **Rise of Digital and Cyber Tools**
   Companies now build not only guns and planes but also cybersecurity solutions. They might develop software to protect military networks or to disrupt an enemy's communications.

2. **Autonomous Systems**
   The future may see more unmanned vehicles—ground robots, drones, and underwater craft—that can perform missions without risking human pilots or drivers.
3. **Commercial Spin-Offs**
   Technology created for weapons sometimes appears in daily life. Radar, the internet, and GPS all began as military projects. This crossover can lead to new business opportunities for defense firms.
4. **Global Competition**
   Emerging powers want to grow their own arms industries, challenging the traditional leaders. This can shake up the market, forcing established companies to innovate to stay ahead.

## Balancing Profits and Responsibility

Many believe that making weapons carries a special responsibility. The items produced can harm or kill people if used aggressively or carelessly. Critics ask if arms manufacturers do enough to ensure their products do not end up with dangerous regimes or criminal groups.

Some companies claim they follow all export laws and carefully screen customers. Others develop ethical policies, pledging not to sell arms to places with severe human rights abuses. However, critics say that as long as there is money to be made, some producers and dealers will look for ways around the rules. This debate about accountability remains heated in the global community.

# CHAPTER 12

## RESEARCH AND DEVELOPMENT IN THE WEAPON INDUSTRY

Creating a new weapon starts long before any parts are assembled. It begins with questions, ideas, and tests—often years of research and development (R&D). In the weapon industry, this process merges science, engineering, and sometimes cutting-edge technology that is still being discovered. Governments and private companies both invest huge amounts of money into making sure their weapons stay ahead of those used by rivals or potential enemies.

In this chapter, we look at how R&D is carried out in the weapon industry, who funds it, and the steps from an initial idea to a functional prototype. We also see how science and design come together to push new frontiers in accuracy, power, and safety. R&D can lead to major breakthroughs, but it can also face obstacles like high costs, technical failures, and political concerns.

## The Importance of Innovation in Weapons

Every major leap in weapon capability—like the move from bows and arrows to firearms, or the shift from simple bombs to guided missiles—happened because someone tested a new idea. This drive for innovation continues today:

1. **Staying Ahead of Threats**: In a fast-changing world, a weapon that is effective today could become outdated tomorrow if enemies develop better defenses or more advanced arms. Constant R&D helps nations and companies keep up.
2. **Filling Gaps**: Sometimes militaries realize they lack a certain capability, such as the ability to strike a target at a longer

range, detect stealth aircraft, or destroy underwater mines safely. R&D aims to solve these challenges.
3. **Improving Efficiency**: Modern armies care about costs, reliability, and performance. A new missile that hits the target with fewer shots can save money and reduce collateral damage.
4. **Export Advantages**: If a country or company can produce highly advanced weapons, they can sell them abroad (if laws allow) and boost their economy. Innovation can become a selling point in the global arms market.

## Funding Sources

Weapon R&D does not come cheap. Often, the process takes years of trial and error before reaching success. Common funding sources include:

1. **Government Defense Budgets**
   Many nations set aside money specifically for weapon research. Agencies within a government may oversee these projects and partner with private companies or public research labs.
2. **Private Investment**
   Some large defense firms use their own profits to invest in new weapon concepts, hoping to gain an edge when the government issues contracts. They may also work with venture capital firms if they have spin-off technologies that could appeal to both military and commercial markets.
3. **International Collaborations**
   Countries sometimes share the financial load by forming alliances or consortiums. For example, several nations might pitch in to develop a next-generation fighter jet. Each country's industry contributes money and expertise, and all get to use the final product.

4. **University Partnerships**
   Universities may receive grants or contracts to do basic research, such as studying materials that can handle extreme temperatures. The results can later be applied to weapons, though this can be controversial if people do not support military work at academic institutions.
5. **Unconventional or Hidden Funds**
   In some cases, R&D might be hidden in secret budgets, especially for projects that are highly classified. Such funding can be harder to track, raising questions about oversight.

## The Research Phase

The first step is often basic research—an exploration of science and technology that might lead to future weapons. It might not have an immediate military goal. Scientists look into new materials, exotic energy sources, or advanced computing methods. Some research is guided by open questions: "How can we see through thick clouds?" or "Is there a way to reduce recoil on large guns?"

During this phase, researchers might:

- Work in laboratories, testing small-scale ideas.
- Use simulations and computer models to predict how an idea might work in real-world conditions.
- Publish or share some findings in scientific journals, though sensitive details could remain classified.

For example, a lab might experiment with a new way to store energy in batteries, not knowing it will eventually help power a laser weapon. Another team might study how super-strong alloys behave under extreme pressure, later discovered to be perfect for tank armor.

## Applied Research and Conceptual Design

Once a promising idea is found, applied research tries to see if it is practical for a real weapon. Engineers and scientists work closely here. They might:

1. **Define Requirements**: If the goal is a new rifle, the design team figures out how accurate it should be, how far it needs to shoot, and how much recoil is acceptable. If it is a submarine-launched drone, they decide how deep it must dive or how quietly it must operate.
2. **Early Prototypes**: Simple models or partial builds to test the concept. These prototypes are often rough, focusing on function more than looks. For instance, an experimental laser might be tested on a lab bench before being shrunk and ruggedized for military use.
3. **Testing and Data Collection**: Teams measure performance against the desired goals. How long does the power source last? Does it work in very hot or cold weather? Tests reveal flaws that must be fixed in the next design.
4. **Risk Assessment**: At this stage, experts evaluate whether the idea is too costly or complicated. They ask if the project should continue, be modified, or be stopped.

---

## Development and Engineering

This phase moves from idea to something approaching a final product:

1. **Detailed Blueprints**: Engineers make precise plans for how each part fits together. This includes 3D computer models, stress calculations, and safety margins.
2. **Working Prototypes**: These are closer to the real weapon. Designers integrate all systems—electronics, software,

propulsion, guidance—so testers can see how it behaves as a whole. These prototypes might be tested in real conditions, such as deserts or arctic environments, to ensure reliability.
3. **Field Trials**: The weapon is used in conditions similar to actual combat. Soldiers or specialized testers try it out. They may identify issues like an awkward grip, a scope that fogs up, or a software glitch that crashes the targeting system.
4. **Design Adjustments**: After each trial, the development team revises the design. Problems found in tests are fixed, new features might be added, or changes are made to simplify production. This cycle can repeat many times.
5. **Cost Analysis**: Throughout, project managers watch the budget, adjusting parts or processes if expenses become too high. Sometimes cheaper materials replace expensive ones, or certain features are dropped.

---

## Prototyping Challenges

Developing new weapons is rarely easy. Designers face many hurdles:

1. **Complex Systems**: Modern weapons can combine mechanics, software, electronics, and sometimes advanced sensors. A glitch in one system can ruin the entire product.
2. **Weight and Size**: There is often a push to make weapons lighter (for soldier-carried equipment) or smaller (to fit in tight spaces). Yet they must remain strong, reliable, and powerful.
3. **User-Friendliness**: Soldiers in the field have to operate these weapons under stress. If controls are too confusing or the weapon is fragile, it might fail when needed most.
4. **Reliability**: A jammed gun or a missile that fails to launch at a critical moment can be disastrous. Developers must design

for harsh conditions—sand, rain, heat, cold—and test for every possibility.
5. **Time Pressure**: Political or military demands can push for quick results. Rushing a design might lead to overlooked flaws, while taking too long can raise costs or miss a window of opportunity.

## Role of Classified Projects

Some weapon R&D is secret. Governments hide details from the public, claiming national security reasons. Classified projects might involve stealth technology, cyber warfare tools, or nuclear research. These programs can be controversial because they often involve large sums of money without public oversight. Also, breakthroughs in secret labs may stay hidden for many years, which can slow civilian scientific progress if the information is not shared.

Still, secrecy can protect critical advantages. If an enemy does not know a new drone or missile exists, they cannot prepare defenses in time. Secrecy can also prevent rival nations from copying or stealing ideas.

## Testing Facilities and Ranges

Prototypes need real-world tests. Weapons are tried out at specialized areas known as test ranges:

- **Shooting Ranges**: For guns and small arms, to measure accuracy, recoil, and reliability over many shots.
- **Missile and Rocket Testing**: Often done in remote deserts or over oceans where a test vehicle can fly without risk to civilians.

- **Naval Testing**: Ships or submarines might test new torpedoes or detection equipment in large bodies of water.
- **Wind Tunnels**: Aircraft designs are tested in scaled form, using powerful fans to simulate flight.
- **Environmental Chambers**: Weapons are subjected to extreme heat, cold, or humidity to see if they still function.

Safety is paramount. If something goes wrong—a rocket veers off course or a new explosive misfires—these remote testing grounds minimize harm. Lessons learned help refine the weapon or, in some cases, end the project if problems cannot be fixed affordably.

---

## Collaboration with Universities and Tech Firms

The weapon industry does not exist in a vacuum. Many breakthroughs come from outside fields like computer science, materials engineering, or robotics. Defense companies regularly collaborate with universities and tech startups to adopt fresh ideas:

1. **Joint Research Projects**: A university might get a grant from a defense agency to study a new metal, while a company provides guidance on the military's needs.
2. **Hiring Students**: Defense firms may recruit graduates skilled in artificial intelligence or advanced math, hoping to harness their talent for future weapon designs.
3. **Innovation Hubs**: Tech companies that develop cutting-edge software or hardware might find their products adapted to military use. This can cause debates about the ethics of using commercial tech for lethal purposes.

This mix of academic research and corporate drive can speed up progress, but it also raises questions about transparency and moral responsibility.

## Software in Modern Weapons

Many advanced weapons rely on software for guidance, targeting, and communication. This includes:

- **Flight Control** for drones, adjusting the craft's position based on sensor data.
- **Guidance Algorithms** for missiles, deciding mid-flight how to correct course.
- **Target Recognition** using machine learning to identify threats from camera feeds.
- **Cybersecurity** measures to prevent hacking.

Writing and testing software is a key part of R&D. A small error in code can cause big failures. Also, software needs to work on rugged hardware that can handle shocks, high temperatures, or electromagnetic interference. This is different from typical consumer electronics, which do not face such stressful conditions.

## Balancing Innovation with Ethics

R&D often pushes the limits of what is possible. But some worry about the ethical side of creating more lethal or autonomous weapons. As technology grows more advanced, it can be used in ways that distance humans from the act of killing. Examples include unmanned drones that strike targets half a world away or AI systems that might one day select targets without direct human input.

Engineers and scientists sometimes question if they want to be involved. Some choose to leave projects if they feel uncomfortable. Others argue that their work helps produce weapons that cause fewer civilian deaths through better precision. Governments and international organizations debate rules on emerging tech, like banning fully autonomous "killer robots." The outcome of these debates can shape future R&D efforts.

## Proving Ground for Civilian Tech

Weapon R&D can lead to inventions that benefit society in peaceful ways. Radar technology, the internet, and GPS were all advanced through military projects before becoming everyday tools. Current research in drones, robotics, and energy storage could bring similar benefits to agriculture, search and rescue, or transportation.

Yet the opposite can also happen. Military needs can drive up prices, limit availability, or keep key patents hidden. Deciding when to release dual-use technology—a product useful for both civilian and military purposes—can be complex, as it might give commercial rivals or potential adversaries a boost.

---

## The Role of Testing Organizations

Before a new weapon is accepted, independent organizations—often part of the military or government—validate its performance. They may run final trials, checking if the weapon meets contract promises. For instance, if a missile is supposed to have a range of 300 miles, the testers see if it actually does in various conditions. If the weapon fails, the manufacturer must correct issues or face penalties.

These testing groups aim to ensure taxpayers' money is well spent and soldiers get reliable gear. Critics warn that if these agencies are not truly independent, they might approve a flawed system due to political pressure or close ties to the manufacturer.

---

## Accelerating R&D with Digital Tools

In recent years, digital methods speed up development:

1. **Simulation and Modeling**: Engineers can run thousands of virtual tests, changing variables such as temperature or wind speed, before ever building a prototype. This saves time and money.
2. **Rapid Prototyping**: 3D printers can quickly create custom parts, letting teams try new ideas without waiting for traditional manufacturing.
3. **Data Analytics**: Using big data, developers can spot patterns in test results, predicting failures or finding ways to improve performance.
4. **Modular Design**: Building weapons with interchangeable parts means developers can swap out sections (like sensors or power supplies) without redesigning the whole system.

These tools reduce the time from concept to testing. However, they do not eliminate the need for real-life trials. A perfect simulation might still miss factors like dust clogging a weapon's moving parts or a soldier's difficulty operating a device in heavy gloves.

## International Cooperation and Competition

R&D can be a race between nations. Countries want the newest technology first, to gain an edge in defense or to sell weapons abroad. This race can lead to large budgets and frantic schedules. At the same time, some projects are cooperative, with multiple nations splitting research costs. Examples include joint fighter jet programs or missile defense systems shared by allies.

Competition can also spark espionage. States or companies might try to steal design secrets, either to replicate them or to catch up faster. Cyber theft of data is an increasing concern, as hackers target labs and defense contractors to acquire valuable information without physically breaking into facilities.

## Challenges Facing R&D

Despite massive investments, weapon R&D faces many challenges:

1. **Escalating Costs**: Advanced materials, specialized testing, and high salaries for skilled engineers can drive budgets to huge levels. Sometimes projects run over budget by large margins.
2. **Political Shifts**: A change in government leadership or policy might cancel or reduce funding for certain programs, stranding years of research.
3. **Technological Limitations**: Even with billions of dollars, some ideas remain out of reach. Building a laser that can destroy missiles reliably in all weather, for example, remains difficult.
4. **Public Backlash**: If people strongly oppose a new weapon—due to cost or moral concerns—political leaders might scrap it.
5. **Export Restrictions**: When a new weapon is made, rules on selling it abroad may reduce potential profits and hamper recouping R&D costs.

---

## Future Directions

Weapon R&D is likely to keep focusing on a few key areas:

1. **Artificial Intelligence**: Smarter guidance, predictive maintenance, and data analysis could streamline everything from target selection to logistics.
2. **Directed Energy**: Lasers or microwave weapons that strike at the speed of light, though challenges remain in generating enough power.
3. **Hypersonic Systems**: Missiles or aircraft traveling at many times the speed of sound, making them hard to intercept.

4. **Swarm Technologies**: Many small, low-cost drones working together can overwhelm defenses through numbers and coordination.
5. **Advanced Materials**: Lighter, stronger alloys or composite materials to improve armor, reduce vehicle weight, and handle extreme conditions.
6. **Enhanced Human Performance**: Exoskeletons, wearable sensors, or biotech enhancements for soldiers to boost endurance and awareness.

As these projects evolve, new safety, ethical, and budget questions will arise. Balancing raw innovation with the risks is part of the ongoing story of weapon R&D.

---

# CHAPTER 13

## MONEY MATTERS – FUNDING AND PROFITS

Weapons do not just appear out of thin air. They come from factories, research labs, and entire organizations that spend time and money to make these products. Behind every tank, jet, and rifle is a long chain of finances—from the first dollar invested to the final payment when a military force receives its order. In this chapter, we look at where the money comes from, how it is spent, and who makes a profit. Understanding these money matters helps us see why the weapon industry is such a major part of the global economy.

## Why Money Drives the Weapon Industry

Producing weapons can be expensive, but it can also pay off. Many countries invest in their defense to protect themselves, keep up with rival nations, or support allies. This demand means companies that build arms can often expect steady contracts. Here are some reasons why money is so important in this field:

1. **High Production Costs**
   Making advanced weaponry—like stealth jets or guided missiles—takes cutting-edge technology, special materials, and lots of skilled workers. These factors push up costs.
2. **Steady Demand**
   Militaries constantly need to replace older equipment or buy new items to stay modern. This can create stable work for the companies involved.
3. **Government Budgets**
   In many countries, the government sets aside a large part of its budget for defense. Some of that money goes directly into contracts with private firms or state-owned manufacturers.

4. **Profit Potential**
   Defense firms can make considerable profits from large contracts. Investors see this as a chance to earn money, making them more likely to back weapon makers financially.

Because of these factors, the weapon industry has become a big part of many economies, influencing job markets, government policies, and even national politics.

## Sources of Weapon Funding

Weapons cost a great deal, so the money to pay for them can come from several places:

1. **Government Budgets**
   The most common source is tax money. Governments collect taxes from citizens and businesses, then allocate part of those funds to defense. Large nations may spend billions or even trillions, which is then divided among different military needs.
2. **Private Capital**
   Some weapon makers are publicly traded on the stock market, so they can raise funds by selling shares. Others get loans or investment from banks, private equity firms, or wealthy individuals.
3. **Foreign Aid or Grants**
   Sometimes a rich country helps a poorer ally by giving or lending money for defense. The ally might use that money to buy weapons, often from the donor nation itself.
4. **International Organizations**
   In rare cases, groups like regional alliances might pool their resources for shared defense projects. Though these budgets are usually smaller, they can be important for countries that cannot afford big programs alone.

5. **Offset Agreements**
   When a nation buys weapons from a foreign maker, the deal may include extra investments or technology transfer. These offsets can lower the net cost for the buyer, but they also require the seller to spend money on factories, training, or local businesses in the buyer's country.

No matter where the money comes from, it usually moves through a complex chain of approvals, contracts, and audits. This ensures that large sums are not misused, though problems still happen.

## Government Defense Budgets

Defense budgets typically cover:

- **Equipment Purchases**: Big-ticket items like warships, tanks, and fighter jets.
- **Maintenance**: Keeping weapons in working condition, buying spare parts, and training crews to operate them.
- **Research and Development**: Funding new technology so the military stays up-to-date.
- **Personnel Costs**: Salaries, housing, and benefits for service members.

Depending on a country's laws, the defense ministry or department might hold the biggest say in how this budget is spent. In democracies, elected officials often vote on these budgets and can ask for changes. In more controlled states, the government might decide with less public input. Either way, the size and direction of a defense budget can shape the entire weapon industry in that nation.

## How Companies Make a Profit

Weapon manufacturers aim to earn money beyond covering their costs. Profits come from:

1. **Sales Contracts**
   Once they sign a deal to deliver a set number of weapons, the company receives payments according to a schedule. Some payments come early for research costs, others arrive when prototypes pass tests, and the rest when the final product is delivered.
2. **Long-Term Support**
   Many deals include services like training operators, providing spare parts, or upgrading systems over time. These can be extra revenue streams for years after the initial sale.
3. **Exports**
   Selling weapons abroad can be very profitable, especially if the company is known for quality. International sales also spread the cost of research across more buyers.
4. **Licensing and Technology Fees**
   If a foreign firm wants to build a weapon under license, the original designer can charge for sharing plans or technical know-how.

By combining these methods, a successful arms maker can become quite profitable. Large companies might also diversify, offering civilian products or security services, making them even stronger financially.

---

## Profit Margins and Investor Interest

Investors keep an eye on profit margins—how much a company earns compared to what it spends. In the weapon industry, margins can be higher than in other sectors, partly because:

1. **Stable Demand**
   Countries rarely stop buying weapons altogether. Many keep upgrading, ensuring a dependable flow of orders.

2. **Long Contracts**
   A fighter jet project might last decades, from initial development to mid-life upgrades. This promises steady income over time.
3. **Complex Products**
   Advanced weapons can be hard to replicate. This exclusivity can lead to less competition and stable prices.
4. **Government Support**
   Some governments grant subsidies or tax benefits to local weapon makers, lowering costs and boosting profits.

Still, these profits must cover massive expenses. Developing next-generation equipment often costs billions, and any technical failure or cancellation can eat into earnings. Investors who back these companies accept the risk in hopes that the big returns from a successful project will outweigh potential losses.

---

## Cost Overruns and Delays

Not everything goes as planned in weapon development. Overruns happen when a project's actual costs exceed the original estimate. Delays can also push up costs, as teams keep getting paid while progress stalls. Common causes include:

1. **Technical Complexity**: Building high-tech gear can reveal unseen problems that require more time and money to solve.
2. **Changing Requirements**: A government might add features or raise standards after the project starts, forcing designers to redo parts of the work.
3. **Supply Chain Woes**: Delays from parts suppliers or raw material shortages can slow production.
4. **Poor Management**: If a company or government fails to plan properly, budgets can spiral out of control.

5. **Unexpected Setbacks**: Natural disasters, economic troubles, or sudden political changes can derail progress.

When overruns happen, the government might increase the budget, or the company might have to absorb the loss, hurting its profits. Sometimes both share the pain, renegotiating the contract to salvage what they can.

## Economic Impact on Communities

Weapon production is not just about global politics or big corporations. It can deeply affect local communities:

- **Job Creation**: Factories and research centers need engineers, technicians, assembly workers, and support staff. Towns near these centers often enjoy low unemployment and decent wages.
- **Training and Skills**: Workers gain specialized skills that can later be used in other industries. In this way, weapon-building can lead to a more trained workforce overall.
- **Local Business Growth**: Suppliers, restaurants, and transportation services may thrive around these factories, boosting the regional economy.
- **Dependency Concerns**: If a weapons program ends, or a contract goes elsewhere, a community might face layoffs and shrinking income. This can harm schools, shops, and public services.

Politicians sometimes push for a share of production to happen in their home region, promising voters more jobs and money. Companies, in turn, highlight how their success benefits everyday people, hoping for continued public support and funding.

## Corruption and Bribery Issues

When large sums are at stake, corruption can follow. Some officials may demand bribes from companies seeking contracts, or a firm might offer gifts to influence the selection process. Investigations in many countries have uncovered shady deals that waste taxpayer money and damage public trust. Common warning signs include:

1. **Secret Agreements**
   Lack of transparency in the bidding process or unusual terms in contracts.
2. **Overpriced Items**
   Costs far higher than normal market rates, with no good reason.
3. **Fake Companies**
   Shell businesses set up to funnel payments or mask illegal activities.
4. **Close Ties**
   Politicians or military leaders with personal or family interests in defense firms.

Many nations have anti-corruption laws, and international treaties aim to limit illegal practices. However, monitoring remains difficult. Companies might claim they only pay "consulting fees," but investigators often see this as hidden bribery. Oversight groups and investigative journalists sometimes expose these problems, leading to reforms or punishments.

---

## Impact of Fluctuating Defense Budgets

Defense budgets can change due to political shifts, economic recessions, or new threats. A sudden increase in spending might happen if a government feels threatened, while a downturn could occur if voters demand less military focus. These swings cause uncertainty for weapon makers:

- **Boom Times**: Firms hurry to fill orders, hire more staff, and raise production. They may invest in new factories or research, hoping demand stays high.
- **Bust Times**: Companies might merge, cut back staff, or seek foreign customers to survive. Small suppliers can go bankrupt if they lose key contracts.
- **Innovation vs. Cost-Cutting**: In lean periods, there is pressure to deliver cheaper options, which can slow down research or lead to simpler, lower-tech weapons.

Some large firms diversify by making civilian products—like passenger planes or electronic systems—so they do not rely solely on defense. Others expand abroad, looking for new markets in different regions.

## Private Equity and Mergers

Private equity firms sometimes buy weapon companies, aiming to increase their profits by restructuring or combining them with others. Mergers occur when two big defense firms unite to form an even larger entity with more resources. This can reduce competition, though, and create "mega-firms" that dominate entire segments. Concerns about these mergers include:

1. **Less Competition**: If only a few giants remain, they could charge higher prices or spend less on innovation.
2. **Political Influence**: A mega-firm might have more lobbying power, shaping defense policy in ways that favor its products.
3. **Worker Layoffs**: Mergers often involve cutting duplicate roles to save money, leaving some employees without jobs.
4. **Dependence**: Governments may become overly reliant on a handful of massive suppliers, giving them little choice when awarding contracts.

On the flip side, a larger firm might afford bigger research projects and manage global supply chains better, potentially improving weapon quality or cutting delivery times.

## Profiting from War or Tension

One of the hardest debates is whether the weapon industry benefits when conflicts arise. More tension often leads governments to spend more on defense. Some critics argue that arms makers have a motive to support policies that keep tensions high. Most companies deny this, saying they simply respond to the needs of their customers. Nonetheless, this issue remains sensitive, as many people do not want to think their safety is tied to someone else's profit motive.

## Transparency and Accountability

To calm concerns about corruption and undue influence, some organizations push for transparency. This can include:

- **Open Bids**: Requiring all major contracts to go through a public bidding process.
- **Public Disclosures**: Publishing contract details, such as the final price and delivery schedule, so taxpayers see how funds are used.
- **Independent Audits**: Outside experts checking if costs are reasonable and the product matches the promised performance.
- **Whistleblower Protections**: Safeguarding employees who reveal wrongdoing, like bribes or fraud.

Such steps do not eliminate all problems, but they can reduce abuse and improve trust among citizens.

## Ethical Investing

Some investors avoid defense stocks entirely, believing that funding weapons goes against their moral values. They prefer "socially responsible" funds that screen out arms makers. Others accept defense companies but draw a line at certain types of weapons, like cluster bombs or landmines, which often affect civilians. This has led to categories of "ethical investing," where individuals or groups only put money into firms that meet specific criteria. As awareness grows, some big investors pressure defense firms to be more open about how their products are used.

## Balancing the Budget with Other Needs

Leaders face tough choices when setting defense budgets. Spending heavily on weapons can mean less money for schools, roads, or healthcare. Governments must decide how to protect their people while also ensuring the welfare of everyday life. Some argue that strong defense is a must in an uncertain world, while others want to see more resources put into diplomacy, economic development, or social services. This balance varies by country, often influenced by historical experiences and perceived threats.

## Shifting Focus to Cyber and High-Tech

As warfare evolves, some of the biggest profits may shift toward sectors like cybersecurity or space-based surveillance. These high-tech areas require large investments in research and specialized talent. Defense companies that adapt quickly might see their revenues rise, especially if governments believe future wars will be fought with information and data as much as with bombs and bullets.

## Global Competition for Defense Dollars

Countries that can produce advanced weapons often compete for export sales. They showcase their products at international arms fairs, where military officials from around the world compare features and prices. Deals can be influenced by political alliances: a buyer might choose a supplier from a friendly nation even if the cost is higher. Meanwhile, rising powers want to develop their own industries and challenge the long-dominant exporters. All this creates a busy global market that moves large sums of money.

---

## Conclusion of Chapter 13

Money is the lifeblood of the weapon industry. From large government budgets to private funds and exports, many different flows of cash keep factories running and research labs busy. In return, the industry offers advanced technology, steady jobs, and often substantial profits for investors and companies. Yet, these gains come with controversies: corruption, moral dilemmas, and the fear that profit might depend on instability or war.

Keeping such vast sums of money in check requires rules, audits, and public debate. Citizens may accept defense costs if they trust their leaders and contractors to act responsibly. Where trust is lacking, concerns grow about wasteful spending or unethical practices. As technology changes and conflicts shift, the balance of spending and saving will remain a hot issue, driving discussions about how much money to invest in arms versus other needs.

In the next chapter, we turn to how these weapons find their way across borders. We will see the methods—both legal and illegal—by which arms flow from factories to far-off lands, and the complicated systems set up to manage these trades.

---

# CHAPTER 14

## HOW WEAPONS ARE SOLD AROUND THE WORLD

Many of the weapons made in one country end up in another, crossing oceans and borders. This global arms trade is huge, involving everything from small firearms to the largest warships. While some deals are carefully approved by governments and follow the law, others happen in secret or break international rules. In this chapter, we will explore how weapons are sold around the world, the different methods of sale, and the rules designed to keep these movements in check.

## The Global Arms Market

The international trade in weapons can be compared to a vast marketplace, but it is not as simple as buying a regular product. It involves:

1. **Government-to-Government Deals**
   Countries often buy directly from another country's official suppliers. Sometimes, this is part of a defense treaty or alliance.
2. **Commercial Sales**
   Private defense firms sell to clients abroad, but they must get permission from their home government.
3. **Illicit Deals**
   Smugglers or black-market traders supply weapons to groups that cannot buy legally, such as certain rebel forces or criminal gangs.

4. **Aid or Gifts**
   A wealthy nation might donate or lend arms to an ally facing threats, usually under strict conditions.

The size of the global arms market shifts each year, but it consistently reaches many billions of dollars in sales. As we saw in the previous chapter, these deals can bring big profits to sellers, while buyers hope the new equipment strengthens their military or security forces.

## Government-to-Government Sales

One common way weapons move across borders is through direct government agreements. In this case, the selling country's armed forces or state-owned factory deals directly with the buyer nation. Sometimes, they arrange:

- **Foreign Military Sales (FMS)**: A system used by some countries, notably the United States, to handle arms exports to allies. The selling government manages much of the paperwork and ensures the equipment meets legal rules.
- **Bilateral Defense Agreements**: Two governments sign a treaty or memorandum saying one will provide weapons, spare parts, or training to the other. Payment might come in cash, or there could be a long-term loan.

Such deals can include more than just hardware. The buyer may want training courses, maintenance support, or technology sharing. These extras can raise the total value of the agreement and lock the buyer into a long relationship with the seller for spare parts and future upgrades.

## Commercial Sales and Export Licenses

Many weapon producers are private companies that want to sell abroad. However, they cannot just ship arms without government permission. Almost every country has export control laws to manage who can buy its weapons. Firms must apply for an export license, detailing:

- **What is being sold**: Rifles, tanks, missiles, or electronic systems.
- **The buyer's identity**: A specific government agency or a private security firm.
- **The final use**: Whether it is for that country's military, police, or resale to a third party.
- **Compliance with Regulations**: Proof they meet all conditions set by the government, including not selling to embargoed nations.

If approved, the license lets the company ship the weapons. This process ensures the home government can veto any deal it finds suspicious or harmful to its foreign policy goals. In practice, many deals do get approved if the buyer is a friendly nation and the weapons are not seen as destabilizing.

## Offsets and Industrial Cooperation

Sometimes, a country buying weapons wants more than just the final product. It might also demand "offsets," meaning the seller has to invest in local industry, share technology, or create jobs in the buyer's country. Examples include:

- **Co-production**: Part of the weapon is built in the buyer's factories, training local workers and transferring some knowledge.
- **Technology Transfer**: The seller shares certain design secrets so the buyer can later maintain or even upgrade the system on its own.

- **Investments in Unrelated Sectors**: In some deals, the seller might fund infrastructure or local businesses to sweeten the agreement.

Offsets can sway the buyer's decision if two weapons are equally good but only one seller offers major benefits to local industry. Critics say offsets drive up costs and reduce transparency, but supporters argue they help developing countries build expertise.

## Arms Fairs and Marketing

Weapon makers promote their products at international arms fairs. At these events, companies display tanks, guns, missiles, and hi-tech gear, often with flashy demonstrations. Military officials from dozens of countries attend to compare features, watch live firing tests (if offered), and discuss potential deals. Some big arms fairs happen annually, drawing attention from media and protest groups.

Marketing in the arms world is more direct than consumer advertising. Sellers rely on personal contacts, official showcases, and behind-closed-doors discussions to reach decision-makers. Brochures might emphasize a weapon's performance, cost, or how it performed in real conflicts. While it is a business environment, these fairs also raise questions about selling items designed to harm people.

## International Laws and Treaties

Several treaties and rules try to control how weapons move across borders:

1. **United Nations Arms Trade Treaty (ATT)**
   Aims to set global standards for regulating the import, export, and transfer of conventional weapons. Countries that joined must assess the risk that the weapons could be used to break human rights laws.

2. **European Union Rules**
   EU members follow a common policy on arms exports, though each nation still has some freedom. The policy bans sales to places under embargoes or where there is a high risk of misuse.
3. **Embargoes and Sanctions**
   Groups like the UN Security Council or regional bodies can forbid selling arms to certain nations engaged in conflict or violating agreements. For instance, if a civil war breaks out, a global arms embargo might try to reduce violence by cutting the supply of new guns and ammo.

Despite these efforts, treaties only work if countries follow them. Some nations do not join certain treaties or choose to interpret rules in ways that allow questionable deals. It is a challenge to enforce these regulations in every corner of the world.

---

# End-User Certificates and Verification

When a company exports weapons, the buyer often signs an **end-user certificate** stating that the items are for its own use and will not be resold or given to third parties without permission. This helps the seller's government track where weapons end up. However, end-user certificates can be faked or ignored. Many cases exist where arms arrived in places not listed on the paperwork, especially in conflict zones.

To reduce such risks, some countries demand post-shipment checks. They send inspectors to confirm that the weapons remain with the original buyer. This can be difficult if the buyer is a foreign military with restricted bases. However, it is one step toward ensuring arms do not travel on to unauthorized groups.

# The Middlemen and Brokers

Sales do not always happen in a clear, direct line from the factory to the final user. Arms brokers act as go-betweens, especially when the buyer and seller are far apart or distrustful. A broker might handle legal details, shipping, translations, or negotiations. While many brokers follow the rules, some engage in shady practices:

- **Secret Transfers**: Hiding the real buyer by sending goods through multiple countries.
- **False Declarations**: Labeling weapon parts as "machine parts" or "farm equipment" to dodge export controls.
- **Payment in Cash or Barter**: Making deals harder to trace. In some conflicts, weapons might be traded for gems, drugs, or natural resources like timber.

Because of this, many countries require arms brokers to be licensed and to submit records of their deals. Enforcing these rules across borders, however, is often a challenge.

---

# Legal vs. Illegal Arms Flows

It can be tricky to draw a line between what is "legal" and "illegal." A government might legally export rifles to a friendly nation's security forces. Later, those forces might lose control of the weapons or sell them illegally. Criminals or insurgent groups can also steal weapons from warehouses or battlefields, creating a secondary market. This black-market trade includes:

1. **Stolen Military Stockpiles**: Guns or missiles taken from poorly guarded depots.
2. **Leftovers from Past Wars**: Some conflict zones are awash with leftover arms that never got collected or destroyed.

3. **Homemade or Modified Guns**: In places with strict import controls, criminals or rebels may craft their own firearms or convert non-lethal items into lethal ones.

Illicit arms flows fuel violence, organized crime, and terrorism. International police agencies like Interpol try to track smuggled weapons, but the sheer number of them around the world poses a major obstacle to peace and safety.

## Big Ticket Items: Ships, Jets, and Tanks

Selling major platforms—like warships, fighter jets, or tanks—is often more complex than dealing in small arms:

- **Long Delivery Times**: Building a submarine can take years, during which the buyer pays in stages.
- **Maintenance Contracts**: The buyer needs training, spare parts, and routine upkeep over the vessel's lifetime. This can lock them into a long-term relationship with the seller.
- **Offset Demands**: Large buys often include demands for local production or tech sharing.
- **Geopolitical Considerations**: Buying a significant platform changes the power balance in a region. Other nations watch closely to see how it might affect alliances or conflicts.

Such deals sometimes spark controversy, especially if the buyer is in a tense region or has questionable human rights records. Critics ask if these arms might escalate local rivalries or be used against civilians.

## Impact of Armed Conflicts on Sales

When tensions rise, arms deals often increase. Nations might rush to build their stocks in case a conflict breaks out. Companies supplying certain equipment—like drones or anti-tank missiles—see higher demand. After a war, leftover weapons can flood the black market, or local armies might want more modern replacements to prepare for future threats. This cycle can keep the global arms business active, as older gear is sold or gifted to smaller states, making room for updated models in wealthier militaries.

## Regional Hotspots and Trade

Certain areas experience heavy arms transfers:

1. **The Middle East**
   Ongoing rivalries and conflicts lead countries here to buy advanced jets, missiles, and defense systems. Wealthy nations with oil riches can afford top-shelf items, while smaller or poorer states may rely on cheaper gear from less restrictive sellers.
2. **Asia-Pacific**
   Rising powers build up navies and air forces, competing with neighbors. Long maritime borders encourage the purchase of submarines and patrol ships.
3. **Africa**
   Many countries face security problems from insurgent groups. Some rely on imports of basic weapons or second-hand equipment. In conflicts, illegal arms often spread across porous borders.
4. **Europe**
   Several nations produce advanced weapons and also import from allies, reflecting membership in alliances like NATO. Some countries shift toward high-tech drones or cybersecurity tools.
5. **Latin America**
   Governments in this region might buy arms to fight drug cartels or modernize older military gear. Illicit flows of small arms also worsen gang violence in parts of Central and South America.

These differences highlight how each region's political landscape affects arms deals.

## Private Security Firms

Not all buyers are national armies. Private security companies might want weapons to protect cargo ships from pirates, guard facilities, or support military operations under contract. Governments sometimes hire these firms to sidestep limits on troop deployments. However, letting private groups acquire heavy arms raises accountability questions. Where do they get their licenses? Can they resell or misuse these weapons? These issues spark debates about the role of private forces in modern conflicts.

## Arms Embargoes and Their Effectiveness

When the international community imposes an arms embargo on a state or a rebel group, it hopes to cut off new supplies of weapons. Yet embargoes are often hard to enforce. Some ways they might fail include:

- **Neighboring States**: Allies of the embargoed party can continue smuggling arms across borders.
- **Hidden Shipments**: Cargo ships might turn off tracking systems, re-flag their vessels under fake registrations, and carry arms into blocked ports.
- **Dual-Use Goods**: Items that can serve both civilian and military purposes, like certain electronics, might be shipped openly, later converted for war use.

Despite these challenges, embargoes do raise the cost of obtaining weapons and can slow conflict escalation. Sometimes they encourage peace talks if one side cannot arm itself as easily.

## Transparency and Tracking Initiatives

To reduce the harmful effects of unchecked arms trading, some groups call for better record-keeping. Proposals include:

1. **Global Arms Registries**: A shared database showing each nation's imports and exports, listing weapon types and destinations.
2. **Marking and Tracing**: Requiring every firearm to have a unique serial number that remains visible even if the weapon is modified, making it easier to trace.
3. **Stricter Customs Checks**: Training border officials to spot hidden shipments, scanning cargo, and using advanced tools to identify suspicious containers.
4. **Cooperation Among Police**: Sharing intelligence between countries so investigators can follow leads across borders.

While some progress has been made, full transparency faces resistance from states that prefer privacy or do not want to reveal their deals. Businesses also worry about sharing data that might help competitors.

---

## Ethical Debates

Selling arms raises ethical questions. If a country sells advanced tanks to a regime with a record of cracking down on citizens, is the seller partly responsible for future abuses? Supporters argue that arms are tools; wrongdoing lies in how they are used. Detractors say some buyers are clearly risky. They urge greater caution or outright bans on sales to governments that break human rights laws.

Non-governmental organizations (NGOs) and peace activists sometimes protest arms fairs or publish reports on questionable deals. This can lead to public pressure. In response, some nations have tightened their export rules to block sales that might be used for oppression or aggression. Yet, big deals often move forward if the strategic or economic gains appear to outweigh moral concerns.

---

## Technological Shifts: Drones and Cyber Weapons

Modern trends in warfare—like the rise of drones or cyber attacks—also shape the arms trade. Unmanned aerial vehicles can be sold in kits that are easier to hide from inspectors. Cyber tools might not even look like weapons, being just pieces of code. This leads to new challenges:

- **Low Visibility**: A program that breaks into enemy computers can be emailed, making it tough for export controls to catch.
- **Rapid Innovation**: Technology evolves fast, so rules become outdated quickly.
- **Ambiguous Use**: A drone might be for mapping farmland or used to drop bombs—hard to tell from the hardware alone.

Regulators are still learning how to handle these novel weapon types. The lines between civilian and military uses often blur, leaving gray areas that can be exploited.

---

## Balancing Security and Commerce

Nations try to balance two things: generating revenue and building diplomatic ties by selling arms, while also preventing dangerous regimes or criminals from getting them. Some governments see exports as a tool of foreign policy, rewarding allies with favorable deals and punishing foes through embargoes. This balancing act can be complicated. A government might sign a big weapons contract with a partner that helps it in some international issue, despite concerns about that partner's human rights record. The money and strategic benefits are deemed more important. Critics call for more consistent policies, warning that short-term gains can lead to long-term problems if those weapons end up fueling violence.

## Conclusion of Chapter 14

Weapons rarely stay within the borders of the country that makes them. Through government-to-government agreements, private commercial deals, or even illegal smuggling, arms pass from hand to hand around the world. This commerce can bring wealth to sellers, boost a buyer's defense, or shift the balance of power in a region. But it also carries risks. If tracking is weak, weapons might slip into conflict zones, worsen violence, or be used against innocent people.

Rules, treaties, and embargoes exist to manage these flows. Yet enforcing them is difficult, especially when profit or power is at stake. People debate the moral duty of arms sellers, whether they should refuse deals that might lead to harm. They also question how to handle new technologies like drones and cyber tools that can cross borders almost unnoticed. As we move forward, balancing legitimate defense needs with global safety remains a huge challenge.

In the next chapters, we will explore how supply chains and distribution channels operate, as well as the government oversight that tries to keep weapons under control. Through this, we will see even more how complicated and far-reaching the weapon industry truly is.

# CHAPTER 15

## SUPPLY CHAINS AND DISTRIBUTION

Weapons are not only complex to design and build—they also need to move safely and efficiently from factories to the people or places where they will be used. This travel from point of origin to final destination is known as the **supply chain**. In the weapon industry, supply chains and distribution networks can become very intricate. They involve different companies, various modes of transportation, and strict security measures. Understanding these pathways helps explain why getting weapons from one location to another can be both costly and time-consuming.

In this chapter, we will look at each step in the supply chain, the ways weapons are shipped, and the challenges that can slow down delivery. We will also see how rules and safety concerns guide every part of the process, from careful packing of ammunition to the final handoff of equipment to the military.

## What Is a Supply Chain?

A **supply chain** is the entire system that moves items from where they are created to where they are used. This can include:

1. **Raw Materials**: Metals, electronics, chemicals, and other base materials must be sourced or extracted.
2. **Manufacturing**: Parts are made in factories, then assembled into finished products, like rifles, missiles, or armored vehicles.
3. **Storage and Handling**: Once built, weapons may be held in secure warehouses or staging areas while awaiting transport.

4. **Transportation**: Ships, planes, trucks, or trains carry the items to their next destination, which might be a port, an airfield, or a military base.
5. **Delivery and Deployment**: Ultimately, the weapons go to the armed forces, police agencies, or other approved groups. Maintenance and spare parts keep them functioning over their lifetime.

In everyday goods, supply chains focus on cost and efficiency. In the weapon industry, security and safety add extra layers of complexity. Mistakes can lead to stolen or lost items that could end up on the black market, or to accidents that harm workers.

---

## Raw Materials and Components

Many weapons depend on specialized metals, plastics, and other substances. For example:

- **High-Grade Steel** for gun barrels.
- **Aluminum or Titanium** for aircraft frames and certain missile parts.
- **Composites** for armor or body panels on vehicles.
- **Electronics** like circuit boards, guidance modules, or radar systems.

Companies making these materials may not always realize they are supplying parts for weapons. Some might produce general-purpose steel or circuit boards that can also be used in civilian applications, known as **dual-use** goods. For certain high-tech materials—like advanced carbon fibers—governments keep track of where they go, to ensure they do not fall into the hands of hostile forces.

Large defense companies often rely on thousands of **suppliers**. These suppliers may be small businesses that specialize in one component, such as gears or sensors. Sometimes, a disruption in the supply of a single vital part can delay a whole production line. For instance, a shortage of a special chip might idle tank assembly if that chip is needed for targeting computers.

## Multi-Tier Supply Chain

Many modern weapons involve so many pieces that their creation is spread among multiple "tiers" of suppliers:

1. **Tier One (Primary Contractors)**: These large manufacturers might build major systems, such as a jet's engine or a missile's guidance section. They coordinate with many smaller suppliers.
2. **Tier Two**: These firms make sub-components for Tier One. If Tier One is building a radar system, Tier Two might produce circuit boards, motors, or specialized software modules.
3. **Tier Three and Beyond**: Further down the chain are smaller workshops and parts makers. They supply raw materials, standard bolts, wiring, or other items that eventually become part of a weapon.

Because of this structure, defense companies must maintain close relationships with their entire network. A single breakdown—like a supplier going out of business or a factory destroyed by natural disaster—can ripple across the chain. Managing this complexity can become a full-time job for supply chain professionals, who track deliveries, costs, and quality at every step.

## Assembly and Testing

Once all the parts arrive at the main factory, weapons are assembled under strict conditions:

- **Assembly Lines**: Some smaller arms, like rifles, can be mass-produced on assembly lines, where workers or robots perform a series of tasks. Quality checks happen along the way to catch flaws early.
- **Module Integration**: Larger systems, such as tanks or aircraft, are often built in modules. For example, one module may hold the engine, while another houses electronics. They are tested separately, then connected into a complete system.
- **Clean Rooms and Special Areas**: For advanced electronics or high-energy devices (like laser systems), a dust-free environment might be needed. Explosive materials could require dedicated buildings to protect workers if something goes wrong.
- **Proof Testing**: Weapons are usually test-fired or powered up in safe areas, such as ballistic ranges or specially shielded rooms, to verify they function as designed. This is crucial for safety and performance.

After a system passes its factory tests, it might be disassembled partially for transport or moved in a fully assembled state if it can fit onto a suitable vehicle.

## Packaging and Labeling

Transporting weapons requires special crates, containers, or racks:

1. **Shock-Proof Containers**: To protect delicate electronics or precise mechanical parts, custom foam or metal frameworks are often used.
2. **Temperature and Moisture Controls**: Some ammunition or missiles can degrade if exposed to high humidity or extreme temperatures. Certain crates are sealed with moisture absorbers or have built-in cooling.

3. **Hazard Labels**: Because explosives or dangerous chemicals are often present, each container must show clear warning labels, following international rules for transporting hazardous goods.
4. **Security Seals**: Containers might have tamper-evident seals. If someone tries to open them, inspectors at the next checkpoint can see that the seal is broken.

Failure to pack items properly can cause serious damage or raise suspicion during inspections. In some cases, countries have detailed regulations on how to label weapons shipments, so border agents know exactly what is inside.

---

## Transportation Methods

Weapons can be shipped by **land**, **sea**, or **air**, depending on size, urgency, and cost:

1. **Trucks and Trains**
    - Good for traveling within a large country or across connected borders.
    - Armored trucks might carry valuable or sensitive items.
    - Trains can move bigger loads (like tanks) but are less flexible in routes.
2. **Cargo Ships**
    - Ideal for large volumes—warships, vehicles, heavy artillery, or containers of rifles and ammunition.
    - Must comply with port security procedures. Some ports have designated areas for military cargo.
    - Longer travel times, but typically cheaper than air freight.
3. **Cargo Planes**
    - Fastest option, often used for urgent deliveries or smaller high-value items (like missiles, electronics, or rare spare parts).

- Requires special handling at airports. Some big cargo planes can carry tanks or helicopters, but capacity is limited and costs are high.
4. **Specialized Transport**
    - Submarines or large naval vessels under their own power. These might sail to the buyer's port or be guided by a crew from the seller.
    - In cases where stealth or secrecy is required, unusual routes or methods might be used, although that carries legal and safety risks.

Due to security concerns, shipments with weapons might travel with armed guards or under military escort. The route can be planned carefully to avoid dangerous zones or places where cargo could be stolen.

---

## Customs and Border Checks

Moving weapons across borders is more complicated than regular goods:

- **Export Licenses**: As discussed in earlier chapters, these documents confirm the seller's government allows the sale.
- **Import Licenses**: The buyer's government also issues permits, approving the arrival of weapons.
- **Transit Permissions**: If cargo passes through a third country, that country's authorities must permit the weapons to go through. Sometimes this is done via maritime or air corridors.
- **Inspections**: Customs officials or special enforcement units may open crates to verify contents match the paperwork. Delays can arise if documentation is incomplete or suspicious.

Because of the dangers posed by smuggling, agencies check for any mismatch between declared cargo and actual items. Large shipments attract attention from intelligence services as well, especially if the route or buyer seems unusual.

## Inventory and Stockpiling

Once weapons reach their destination, they often go into storage before actual use:

1. **Military Depots**: Large bases or depots where armies keep spare tanks, artillery, or munitions. These sites have fences, guards, surveillance cameras, and sometimes advanced sensors to detect tampering.
2. **Armories**: Smaller storage for small arms and ammunition, typically with thick walls, locked doors, and strict sign-in/out logs.
3. **Rotation and Maintenance**: Weapons do not just sit there; staff must check them regularly. Ammunition can expire or degrade, so older stock is used first. Tanks or trucks must be started up, or their batteries might fail.
4. **Record-Keeping**: Detailed logs track each weapon's serial number and location. If something goes missing, the staff must investigate immediately.

By maintaining a careful stockpile system, militaries can deploy equipment quickly, knowing exactly where each item is. Poorly managed depots risk theft, accidents, or confusion during an emergency.

## Distribution to End Users

The final step is handing out weapons to the units or teams that will operate them:

- **Military Units**: Soldiers might receive rifles, gear, and vehicles from the unit's armory. The process includes checking out each item by serial number.
- **Police Forces**: Law enforcement agencies have their own supply chain for pistols, riot gear, and specialty weapons if needed.
- **Private Security**: In places where private security is allowed heavier arms, they must store them in authorized locations and maintain records.
- **Allied Forces**: Sometimes a country might lend or lease weapons to a friendly nation under special agreements. This involves transferring equipment and training those who will use it.

If these weapons move on to another force or are sold second-hand, a new supply chain process begins—more paperwork, shipping, and storage.

## Challenges and Delays

Supply chains for weapons face specific hurdles:

1. **Security Threats**: Armed groups might target convoys carrying valuable equipment, especially in conflict areas. Piracy at sea is also a worry for ships carrying arms.
2. **Complex Paperwork**: One missing approval can halt a shipment at a border for days or weeks. The longer the wait, the higher the storage and security costs.
3. **Logistical Bottlenecks**: Ports can get congested, especially if large volumes of cargo arrive at once. Some specialized items (like certain types of explosive) may only be handled at specific terminals with the right safety standards.
4. **Changing Requirements**: A government might shift its needs mid-delivery, requesting different variants of a vehicle or

updated electronics. Adjusting on the fly can disrupt production schedules.
5. **Global Events**: Natural disasters, pandemics, or political instability can shut down key transport routes. When a railway or port is out of commission, alternative routes might be longer and costlier.

All these factors increase the cost and effort to move weapons effectively. Big players in the arms market often set up entire logistics departments dedicated to anticipating and solving these problems.

## The Role of Freight Forwarders

A **freight forwarder** is a company that organizes shipments for others, whether it is consumer goods or specialized freight like weapons. They can handle:

- **Documentation**: Getting the correct licenses, customs papers, insurance forms, and handling instructions.
- **Route Planning**: Choosing a safe and quick route, booking cargo space on ships, planes, or trucks.
- **Consolidation**: Grouping smaller shipments together if that is more cost-effective, though with weapons, consolidation is trickier due to safety concerns.
- **Tracking**: Using software and contacts to update the client on the cargo's location.

Some freight forwarders specialize in handling dangerous or military cargo, knowing the ins and outs of regulations. Hiring such experts can reduce a manufacturer's headaches, but it adds an extra link in the chain where mistakes or misconduct could occur if the forwarder is not reputable.

## Digital Tools and Tracking

To handle complex chains, the weapon industry has adopted digital systems:

1. **Enterprise Resource Planning (ERP)**: Large defense firms use ERP software to track orders, inventory, and finances in real time.
2. **Blockchain or Secure Databases**: Some experiments use blockchain-like technology to maintain tamper-proof records of each item's journey, from raw material to final use.
3. **GPS and IoT Devices**: Sensors on crates can report temperature, humidity, or if a crate was opened. GPS trackers can alert the owner if a truck strays from its planned route.
4. **Automated Alerts**: If shipping times exceed set limits or a container's seal is broken, the system flags it for immediate investigation.

These innovations aim to improve transparency and reduce the chance of lost or stolen goods, though they come with added costs and the need for reliable internet or satellite coverage.

---

## Handling Sensitive Components

Certain parts of a weapon—like the guidance system in a missile or advanced targeting software—may require extra steps:

- **Separate Shipments**: To reduce risk, a missile's warhead might travel apart from its guidance unit, meeting only at a secure facility.
- **Escort by Security Forces**: Governments may insist that specific high-tech items move under armed guard.
- **Secure Warehousing**: Facilities with biometric entry controls, advanced alarms, and constant monitoring may be used for extremely sensitive systems.

This reduces the likelihood that criminals, enemy agents, or terrorists can intercept a critical piece of technology.

## Environmental and Safety Concerns

Transporting weapons can pose environmental risks:

- **Accidents**: A truck carrying explosives could crash, causing blasts or chemical spills. In maritime transport, a sinking ship might spill dangerous materials into the ocean.
- **Improper Disposal**: If weapons reach the end of their service life, they must be disposed of carefully. Some chemical or biological agents are especially hazardous if not destroyed under strict protocols.
- **Fuel Use and Pollution**: Large cargo planes and ships burn significant amounts of fuel, contributing to emissions. Though this is not unique to weapons, it is part of the overall environmental footprint of defense logistics.

Governments and companies often have emergency response plans for accidents. They rehearse steps to contain fires, explosions, or pollution. High-level safety standards help minimize these dangers.

## Illegal Diversion and Smuggling

Despite strict rules, some weapons get diverted or stolen. Smugglers may bribe port officials, forge papers, or hide arms in legal cargo. A cargo container labeled "machinery parts" might in fact hold rifles or missile components. Once these items disappear from legitimate channels, they can surface in conflict zones or with criminal organizations.

Combatting illegal diversion relies on intelligence work, random inspections, and cooperation among countries' law enforcement. Tracking known smugglers or shady shipping companies also helps. Yet, no system is perfect, and the global scale of maritime and air traffic offers many chances for criminals to conceal illicit shipments.

## Wartime Logistics

In active war zones, supply chains take on a more urgent character:

- **Frontline Delivery**: Trucks or aircraft might face enemy fire or sabotage. Delivering ammunition or spare parts to a unit under siege can be extremely risky.
- **Mobile Depots**: Militaries set up temporary warehouses close to the action, moving them as battle lines shift.
- **Destroyed Infrastructure**: Bridges, roads, and rail lines might be damaged, forcing use of alternative routes like air drops or convoys through unsafe regions.
- **Allied Cooperation**: Partner nations might share cargo planes or coordinate deliveries, trying to keep each other supplied. This can be a lifeline in extended conflicts.

The complexity grows even more if multiple allied forces operate together, each with different procedures and equipment needs.

## Relationship with Private and Civilian Sectors

Military supply chains frequently rely on civilian transport companies, shipping lines, and airports. In peacetime, these relationships are governed by contracts and standard laws. During a crisis, governments can invoke emergency powers to prioritize or requisition civilian vehicles, a practice that sometimes causes public debate. If cargo planes meant for general commercial use are diverted for urgent defense needs, it can disrupt civilian trade.

## Training and Workforce

Managing arms logistics calls for specialized workers:

- **Logistics Officers** in the military: Plan shipments, arrange safe storage, and coordinate with civilian authorities.
- **Supply Chain Managers** in defense firms: Oversee supplier contracts, track inventory, and handle shipping details.
- **Customs Experts**: Understand the laws, tariffs, and licenses needed for cross-border arms movement.
- **Warehouse and Port Staff**: Handle packaging, loading, and record-keeping. Must be trained in the dangers of explosive or classified items.

This workforce must remain up-to-date on regulations, new security threats, and evolving technology, since mistakes can lead to huge consequences.

## Future Trends in Weapon Logistics

1. **Automation**: More robotic systems for warehouses or driverless trucks could lower labor costs and reduce human error.
2. **Green Initiatives**: Pressure to cut carbon footprints might lead to more efficient shipping, eco-friendly fuels, or advanced vessel designs.
3. **Data Analytics**: Real-time tracking and predictive analytics could spot potential supply chain disruptions before they happen.
4. **Advanced Manufacturing**: With 3D printing, some spare parts might be created on-site, cutting shipping costs and wait times.

5. **Increased Security Focus**: As criminals and rogue groups become more sophisticated, new scanning or detection tools will likely emerge for cargo screening.

These developments aim to make the supply chain faster, safer, and more transparent. Yet they will still have to align with strict government policies and face the ever-present risk of diversion or theft.

## Conclusion of Chapter 15

From raw materials to the final user, the supply chain in the weapon industry is a complicated process packed with security checks, regulatory hurdles, and the constant need for precision. Every step, from forging metal parts to handing a rifle to a soldier, must be done with care to avoid accidents, misplacements, or illegal diversions. This level of complexity means the industry relies on many experts: engineers, managers, freight forwarders, and customs officials, all working under tight rules.

Despite careful planning, delays and cost overruns can still happen. Interruptions in supply lines can halt entire projects, while misrouted cargo could wind up in the wrong hands, intensifying violence somewhere in the world. That is why the weapon supply chain is subject to many layers of oversight—both by the companies themselves and by the governments that regulate and purchase these products.

In the next chapter, we will take a closer look at how governments create rules and maintain oversight to manage these risks, keep track of sales, and ensure that the weapons do not end up in places or with groups they should not. This government role is vital for maintaining peace and stability, at least in theory, as it sets boundaries for which deals are allowed and how each weapon must be monitored after it is sold.

# CHAPTER 16

## GOVERNMENT RULES AND OVERSIGHT

Governments have a powerful say in who can make weapons, who can buy them, and how they are used. Without official approval, a manufacturer might never get the materials it needs, or an import deal could collapse. Laws, regulations, and oversight bodies form a web of controls around the weapon industry. These measures try to ensure that arms do not land in the wrong hands, that companies follow safety standards, and that the public can trust how their tax money is spent on defense.

In this chapter, we look at how governments develop rules, which agencies enforce them, and the methods used to oversee weapons from the moment they are planned to the day they are retired or destroyed. We will see that while such oversight can be strict, it is not perfect. Conflicts, corruption, and politics all shape how well these regulations work in reality.

## Why Governments Regulate Weapons

Weapons are not just regular products. Their misuse can cause large-scale harm, threaten national security, or lead to human rights abuses. Because of this, governments use laws and policies to manage aspects of the weapon industry:

1.  **Preventing Misuse**: By limiting who can own or produce arms, they hope to reduce crime, terrorism, or internal unrest.

2. **International Commitments**: Countries sign treaties that require them to control exports of certain weapons or technologies.
3. **Security of Supply**: In some places, the government wants to ensure a stable local industry, so they do not rely on foreign suppliers. This can lead to rules that protect home-grown defense firms.
4. **Economic Management**: Defense spending can be huge. Oversight aims to keep corruption low and spending efficient so that taxpayers are not cheated.
5. **Political Influence**: Controlling arms exports can serve foreign policy. A state might reward allies with easier access to weapons or punish foes by banning sales to them.

Each government sets policies based on its own history, threats, and alliances. Even countries with similar values, like members of certain international organizations, can have very different approaches to regulating weapons.

## National Laws and Agencies

Almost every country has laws that detail how weapons can be produced, sold, and used. These might include:

- **Licensing Requirements**: Companies must obtain permits to build or export arms. Officials check for the firm's trustworthiness, technical ability, and financial soundness.
- **Background Checks**: Individuals or businesses seeking to buy certain weapons must prove they have no criminal ties or suspicious motives.
- **Limits on Military-Grade Arms**: Civilians cannot legally buy tanks, fighter jets, or other advanced systems in most places. The same goes for certain destructive items like heavy explosives.

- **Marking and Tracking**: Guns or missiles might need unique serial numbers so the government can trace them if they are used in a crime or discovered in illegal trade.

Enforcement often falls to specific agencies. In some nations, a "Ministry of Defense" works with a "Ministry of Commerce" or a specialized export control authority. Police, customs offices, and border patrol all play parts in preventing illegal movement of arms. Sometimes, intelligence agencies watch for red flags, like suspicious financial transactions or shipments disguised as something else.

## Oversight in Production

For manufacturers, government oversight usually starts early:

1. **Factory Inspections**: Officials check if the company has proper security, such as fences, alarms, and restricted access areas. They also ensure safe handling of explosives or chemicals.
2. **Technical Audits**: Defense agencies might review blueprints, safety tests, and quality control processes. For highly classified items, certain staff must hold security clearances, limiting who can work on the project.
3. **Contracts and Milestones**: When a government funds a project, it sets goals and timelines. If the company fails to meet them, the government can reduce payments or cancel the contract.
4. **Record-Keeping Requirements**: Manufacturers have to track all materials received and weapons produced. Auditors can match these records against actual inventory.

These steps, in theory, guard against theft, accidents, and poor-quality products. However, large companies with political ties or vast influence might, at times, dodge thorough oversight. Smaller firms, meanwhile, can struggle with the cost of compliance—paying for secure facilities or specialized staff.

# Export Control Systems

**Export control** is about deciding which weapons or technologies can leave the country and where they can go. Governments often divide items into categories. For instance:

- **Military List**: Tanks, fighter jets, ammunition, missiles, and other items made specifically for combat.
- **Dual-Use List**: Technologies or products with both civilian and military uses, like certain software, electronics, or chemicals.
- **Nuclear-Related**: Materials and components used to build nuclear weapons or reactors, often under extremely tight control.
- **Sensitive Technologies**: Advanced research that could boost an adversary's weapons program if shared.

Export control laws require companies to apply for licenses before shipping these goods. Officials check:

1. **The Destination**: Is the buyer's country friendly or under embargo?
2. **End-User**: Is it a government defense ministry or a private group with potential links to militants?
3. **Intended Use**: Will this technology support a peaceful satellite launch, or could it lead to ballistic missile development?

Agencies might consult intelligence data or foreign policy departments. Approvals often take time, and some deals face denial if they risk fueling instability or clashing with alliances.

# Import Regulations

On the **import side**, governments set policies about what weapons can enter and who can bring them in. This can include:

- **Permits for Specific Weapons**: A country might allow assault rifles for its police but ban them for civilians, or only let the military import tanks.
- **Testing and Certification**: Imported items may need to pass local safety tests or meet certain performance standards.
- **Tax and Duty Exemptions**: If the items go to the national military, they might be exempt from certain taxes. Private buyers could face high tariffs or fees.
- **Quotas or Bans**: Some nations limit how many firearms can be imported each year, or ban entire categories like handguns.

These rules reflect a government's approach to public safety, national security, and foreign relations. They also shape how big the local arms market can become.

## Budget and Spending Oversight

For weapons bought by the state, **public spending oversight** is crucial:

1. **Parliamentary or Congressional Reviews**: In many democratic countries, elected bodies must approve large defense budgets or key purchases. They can hold hearings, question military officers or industry representatives, and request audits.
2. **Inspector Generals and Watchdogs**: Some governments have independent agencies that track how money is spent. They might check if the contract was awarded fairly or if any part of the budget is misused.
3. **Public Disclosure**: Details about major contracts or budgets might be released to the public, though certain aspects could stay classified for security reasons.

4. **Anti-Corruption Measures**: Officials must declare conflicts of interest. Companies that are caught bribing can be fined, barred from future contracts, or see their executives face legal charges.

When large sums of money flow into complex programs (like developing a stealth aircraft), cost overruns and delays can occur. Proper oversight tries to catch these issues early. Still, critics say that lobbying and political pressure sometimes sway decisions, leading to expensive outcomes that benefit certain businesses over the taxpayer or the armed forces themselves.

## Military End-Use Checks

A key aspect of government oversight is making sure arms do not go to the wrong user:

1. **End-User Certificates**: As mentioned before, these legally bind the buyer to keep the weapons for their stated purpose.
2. **Post-Delivery Inspections**: Some exporting countries send officials to the buyer's facilities to confirm the items remain where they are supposed to be.
3. **Information Sharing**: Governments might alert each other if they find a shipment diverted or suspect black-market activity.
4. **Condition on Re-Exports**: A seller might forbid the buyer from re-selling the weapons to a third party without prior consent. This helps the original manufacturer's government maintain control.

However, these checks only work if the buyer cooperates. In conflicts where governments collapse or lose track of their arsenals, oversight essentially disappears. That is why many arms end up in unexpected places despite official controls.

# Arms Embargoes and Sanctions Enforcement

When a government or an international body imposes an arms embargo, it bans or restricts arms transfers to certain destinations. Enforcement usually involves:

- **Border and Customs Controls**: Extra screening for shipments going to the embargoed location.
- **Blacklist of Entities**: Companies or individuals known to supply banned parties can be added to a blacklist, making it illegal to do business with them.
- **Satellite and Naval Patrols**: Naval forces or surveillance satellites may watch shipping routes to intercept suspicious vessels.
- **International Cooperation**: Since smugglers can reroute cargo, multiple countries must coordinate to close loopholes.

Still, embargoes can fail if major powers do not agree or if neighboring countries ignore the restrictions. Some nations have been accused of secret arms deals that violate sanctions for strategic or economic gain. That is why embargoes, while symbolic, are not always effective in halting the flow of weapons.

---

# Transparency Measures

To prevent illegal or unethical sales, some governments and organizations push for more **transparency**:

1. **Arms Trade Reports**: Publishing the quantity and type of weapons exported, along with the buyer's identity.
2. **Public Hearings**: For large deals, politicians might hold hearings where defense officials explain the rationale.
3. **Company Disclosures**: Firms in some countries must list who they have sold weapons to, unless it is classified.

4. **Monitoring NGOs**: Non-governmental organizations track shipments, contract awards, or suspicious activities, sharing their findings publicly.

These steps let the public and global community spot patterns, such as repeated sales to known human-rights violators, which can spur political or social backlash.

## Licensing and Monitoring Private Security

Not all armed parties are official militaries. Private security companies and contractors sometimes carry weapons for guarding ships, embassies, or infrastructure in dangerous areas. Governments must license these firms, deciding which weapons they can own or transport. They might also watch how these firms store guns and handle training:

- **Background Checks on Personnel**: Ensuring employees do not have serious criminal records or extremist links.
- **Use-of-Force Guidelines**: Setting out when private guards can fire their weapons.
- **Regular Audits**: Checking the company's inventory to confirm no firearms are lost or sold illegally.

Missteps in this area can cause international incidents—if private contractors abuse their weapons abroad, their home country might face diplomatic fallout.

## Nuclear and Other Mass-Destruction Arms Oversight

For **nuclear, chemical, or biological weapons**, oversight is even stricter, often shaped by global treaties:

1. **Nuclear Non-Proliferation Treaty (NPT)**: Nations agree not to share nuclear weapons with non-nuclear states and to only use nuclear power for peaceful goals.
2. **International Atomic Energy Agency (IAEA)**: Inspects nuclear facilities to ensure they are not producing weapons-grade material.
3. **Chemical Weapons Convention**: Bans chemical arms and requires destruction of existing stockpiles under international supervision.
4. **Biological Weapons Convention**: Prohibits developing or producing biological weapons, though verification is harder than in chemical or nuclear fields.

Countries holding such weapons are subject to heavy scrutiny. If a state tries to hide a secret program, other nations can impose sanctions or even threaten military action. This high-stakes situation makes oversight and verification crucial to prevent the spread of extremely dangerous arsenals.

## Audits and Investigations

When something goes wrong—like a stolen shipment, a bribery scandal, or a contract that ballooned in cost—governments may launch **investigations**:

- **Parliamentary Inquiries**: Politicians question officials and company executives in public sessions.
- **Criminal Probes**: Police or special prosecutors look for evidence of corruption or illegal activity.
- **Commissions or Panels**: Independent experts review the situation and recommend reforms.
- **Company Sanctions**: Firms found guilty might lose their licenses, face heavy fines, or be barred from future contracts.

High-profile cases often draw media attention, spurring calls for stricter rules. As a result, some nations periodically update their arms control laws in reaction to scandals.

## The Role of International Organizations

Organizations such as the **United Nations**, **European Union**, or **African Union** can set region-wide or global standards. They pass resolutions on arms control, sponsor peacekeeping efforts, and encourage member states to share data. However, these bodies rely on nations to enforce the rules. If a powerful state chooses not to cooperate, the organizations have limited power to force compliance.

Other groups, like the **Wassenaar Arrangement**, bring together countries to coordinate export controls on conventional arms and dual-use goods. These voluntary bodies help members compare best practices, maintain updated lists of restricted items, and watch for suspicious developments. While not as binding as a treaty, such cooperation can increase pressure on states to keep their arms trade in check.

## Balancing National Security and Commerce

Governments face a balancing act: they want to support local industry and keep strong militaries, but also must block dangerous transfers and corruption. Too many rules might stifle competition or push the arms business underground. Too few rules risk fueling conflicts or scandal. Some governments tilt toward leniency, seeing arms exports as a valuable source of income and diplomatic leverage. Others maintain tight controls, especially if their publics are sensitive to global conflicts or human rights concerns.

# Criticisms of Government Oversight

Despite official frameworks, critics argue that:

1. **Loopholes Remain**: Companies exploit vague wording or complex corporate structures to move weapons around.
2. **Lack of Transparency**: Many deals remain secret under the label of "national security." This can hide corruption or questionable sales.
3. **Inequality**: Powerful nations set rules that smaller states must follow, but do not always apply the same standards to their own arms deals.
4. **Politicized Decisions**: Allies get easy approvals, while the same weapons might be denied to others for political reasons rather than purely moral or security grounds.

Activists and some politicians push for reforms, seeking more consistent and open approaches. They note that oversight usually works best when countries have strong legal institutions, free media, and engaged citizens.

---

# Technology's Impact on Oversight

Emerging tech changes how governments regulate:

1. **Cyber Warfare Tools**: Hard to classify under traditional arms laws, yet can be very destructive. Some states impose separate export rules on hacking software or surveillance tech.
2. **3D Printing**: Potentially allows the local creation of weapon parts from digital files, bypassing normal shipping controls.
3. **Autonomous Systems**: Drones or robots that can make targeting decisions with minimal human input raise moral and legal dilemmas. Governments struggle to define how to oversee or limit them.

4. **AI-Enhanced Weapons**: If software can adapt or learn, who ensures it remains within lawful use? This question continues to grow as artificial intelligence improves.

Agencies must keep updating regulations, but they often move slower than technology does, leaving gaps for misuse.

## Success Stories and Ongoing Struggles

Government oversight has seen real achievements:

- **Reduction of Nuclear Arsenals**: Treaties like START (Strategic Arms Reduction Treaty) cut the number of deployed nuclear warheads in the U.S. and Russia.
- **Chemical Weapons Destruction**: Many nations have destroyed their chemical stockpiles under global agreements.
- **Bans on Landmines and Cluster Munitions**: Although not universal, some treaties have encouraged many countries to stop using or producing these weapons, helping reduce injuries to civilians.

Yet, wars and human rights abuses continue in many places, some fueled by arms that slip through the oversight cracks. Smuggling, bribery, and shifting alliances all hamper strict enforcement.

## Looking Ahead

Governments are likely to face new pressures:

1. **Public Opinion**: As information spreads faster, citizens can apply rapid pressure through social media or other channels when they learn about questionable arms deals.

2. **Global Power Shifts**: Emerging nations might develop their own major arms industries, leading to new export players who set different rules.
3. **Climate and Resource Conflicts**: Changing environmental conditions might spark new unrest, increasing demand for arms and complicating oversight.
4. **Technological Breakthroughs**: Directed energy weapons, quantum computing, or other breakthroughs could outpace regulations, forcing a constant race between rulemakers and innovators.

The core idea remains: weapons are special goods that can safeguard a country or inflict great harm. Governments will keep trying to walk that line between enabling defense and preventing chaos.

---

# CHAPTER 17

## SAFETY AND SECURITY CONCERNS

Weapons can protect people, but they can also pose risks if mishandled, stored poorly, or used carelessly. Even well-run militaries and police forces face safety challenges, and any mistake can cause tragic accidents or put powerful tools in the wrong hands. Security concerns are just as pressing. Stolen weapons, cyber attacks on weapon systems, and insider threats can all lead to severe problems for governments, companies, and citizens. In this chapter, we look at the many safety and security issues surrounding weapons, the ways to address these problems, and why it remains an ongoing challenge.

## The Importance of Safe Handling

Weapons are meant to project force. But if someone does not know how to manage that force, harm can happen unexpectedly. Safe handling aims to stop accidental discharge, unintentional injury, or damage to equipment. Some key aspects include:

1. **Proper Training**
   People who use firearms, tanks, or advanced systems must learn the basics: how to load and unload correctly, maintain muzzle discipline (never pointing a gun at something they do not want to harm), and check for ammunition. Mistakes often occur when someone assumes a gun is empty or forgets a step. In militaries, training is continuous. In civilian settings, courses may be required by law or strongly encouraged.
2. **Clear Procedures**
   Detailed rules cover everything from how to carry rifles during patrols to how to power down a missile launcher. When these steps are followed exactly, the chance of accidents goes down.

3. **Safe Storage**
   Storing weapons separately from ammunition, using locked cabinets or special safes, and tracking who has the keys all reduce the risk of unauthorized access or accidents. In large depots, multiple locks and daily logs of who enters ensure better control.
4. **Regular Maintenance**
   A jammed mechanism can lead to unexpected misfires, so scheduled checks help spot worn parts, dirt buildup, or other hazards. This is especially vital for items that rely on electronics or hydraulics, such as modern artillery or aircraft weapons.

Safe handling may sound simple, but it relies on discipline. Human error or complacency can undo the best protocols. This is why repeated practice, drills, and safety culture matter so much in any organization that uses weapons.

---

## Accidents in Military and Civilian Life

Despite precautions, accidents do happen with severe consequences:

- **Training Mishaps**: Live-fire exercises, where soldiers practice shooting real ammunition, can lead to injury if someone stands in the wrong place or a bullet ricochets.
- **Explosive Handling Errors**: Bombs, mines, or grenades contain materials that can ignite or detonate if dropped, exposed to static electricity, or handled incorrectly.
- **Air or Naval Incidents**: Aircraft carrying missiles may crash, or naval vessels might have accidents with torpedoes or onboard guns. Even minor slip-ups can escalate because of the destructive nature of the cargo.

- **Civilian Gun Accidents**: In places where private gun ownership is common, accidental shootings—especially involving children—get wide attention. Often, these stem from loaded weapons not being locked away.

Some militaries release detailed incident reports so others can learn from their mistakes. Over time, better procedures reduce some dangers, but the human factor remains unpredictable. For instance, a tired or stressed worker might overlook a basic safety rule. That is why leadership tries to maintain strict routines and watch for signs of burnout or rule-breaking.

---

## Storage and Stockpile Safety

National stockpiles can be huge. Armies keep large amounts of ammunition, spare parts, and older weapons. If stored poorly, these can become ticking hazards:

1. **Aging Munitions**: Over decades, explosives inside shells or bombs can degrade, making them more unstable. Temperatures or humidity can speed this decay. A single spark or shock in a badly maintained warehouse might set off chain reactions.
2. **Fire Hazards**: Warehouses that lack sprinklers or firebreaks put entire cities at risk if a blaze starts.
3. **Leaking Chemicals**: Certain weapons use toxic or corrosive substances. If containers fail, these chemicals can harm the environment or sicken workers.
4. **Stockpile Explosions**: History has shown examples of massive blasts in depots that destroy nearby neighborhoods. Investigations often find that improper handling, poor ventilation, or ignoring official guidelines were key factors.

To prevent disaster, many countries follow international best practices on storing munitions. This includes spacing out different types of explosives, using bunkers designed to direct blasts upward rather than outward, and rotating stock so older items get used or safely dismantled first.

## Transportation Risks

When weapons are on the move—by truck, ship, or plane—accidents can be even harder to control. A crash, rough seas, or runway mishap might harm the cargo. If explosives or sensitive gear spill, the local area can become contaminated, or criminals might snatch items during the chaos. Secure transport procedures often include:

- **Specialized Vehicles**: Armored trucks or sealed containers with shock absorbers and temperature controls.
- **Trained Crews**: Drivers and handlers who know how to respond if a box tips over or a leak is suspected.
- **Emergency Plans**: Clear instructions on what to do if an accident occurs, including contacting authorities, sealing off roads, and calling hazardous material teams.

Sometimes, militaries arrange armed escorts to deter hijacking. In conflict areas or pirate-prone waters, a security team might sail alongside the cargo ship. This raises costs but can save lives and prevent stolen goods from fueling further violence.

## Human Error and Fatigue

Weapons demand respect from the people who operate them. However, human factors such as fatigue, stress, or complacency can undo even the best training:

1. **Stress in Combat**: Under enemy fire, a soldier might forget steps, leading to a misfire or dropping a loaded weapon. Adrenaline and confusion are hard to manage.
2. **Routine Boredom**: In peacetime, guards or factory staff might become sloppy with checks, especially if they rarely see real danger. Complacency grows over time if no incident has happened recently.
3. **Shift Work**: Factories often run long hours. Tired workers on night shifts might overlook a loose bolt or store chemicals incorrectly.
4. **Pressure to Deliver**: Managers could push employees to rush assembly or skip safety steps to meet deadlines, raising the chance of mistakes.

Military leaders and company managers try to address these risks through shift rotations, mandatory breaks, mental health support, and random audits to verify everyone follows safety rules. Still, no system is foolproof.

## Stolen or Lost Weapons

One major security concern is theft. Criminals, insurgent groups, or dishonest insiders might try to steal guns, missiles, or crucial parts. Stolen weapons can move quickly into underground markets, making them near-impossible to trace. Key vulnerabilities include:

- **Insider Threats**: Someone working in a depot or factory might smuggle out pieces of a rifle over several days or tamper with shipment records to hide missing goods.
- **Armed Robbery or Break-Ins**: Although weapon sites are guarded, a determined group might attack a poorly staffed checkpoint.

- **Shipment Hijacking**: Road convoys can be ambushed, especially in remote or conflict-ridden regions. Thieves may also bribe transport staff to turn a blind eye.
- **Ghost Guns**: In some places, criminals assemble firearms from unregulated parts. They may only need a stolen or illegally purchased key component, like a barrel or receiver, to build a functional gun.

To combat theft, security systems use cameras, motion detectors, ID checks, and random searches of employees' bags. Electronic inventory systems can flag discrepancies in real time. Yet, smugglers continually find ways to bypass these defenses, especially if corruption seeps into the organization.

## Digital Security Threats

Modern weapons rely heavily on computer networks. Missiles have software-based guidance, tanks have digital controls, and entire command systems connect to satellites. These digital platforms come with new security problems:

1. **Hacking and Malware**: A hacker could disable a weapon system, steal design files, or trick a guidance system into targeting the wrong location.
2. **Supply Chain Attacks**: Malicious code might be inserted into a component's software at the factory, lying dormant until activated.
3. **Electronic Warfare**: Opponents might jam signals or feed false data to drones, causing them to crash or return to base.
4. **Sabotage**: Insiders with programming access can alter code, potentially causing catastrophic failure during operations.

Defense groups invest in cybersecurity. They encrypt data, isolate critical networks from the internet, and regularly scan for intrusions. However, advanced hackers—sometimes backed by rival governments—remain a serious threat. Even minor programming flaws can create entry points for cybercriminals.

## Insider Threats

Not all security risks come from external criminals. An insider threat is someone within the organization who deliberately abuses their access. This could be:

- **A Disgruntled Employee**: Feeling unfairly treated, they might plan sabotage or leak secrets to outsiders for revenge.
- **Ideological Turncoat**: Convinced by extremist beliefs, they aim to arm a militant group or carry out an attack from the inside.
- **Financial Motivation**: A person in dire need of money could sell classified info or physically steal items.
- **Blackmail**: Spies might threaten an employee or trick them into believing they have no choice but to cooperate.

Organizations fight insider threats by screening employees thoroughly, limiting who can enter certain rooms, and monitoring unusual behavior—such as accessing files they do not normally use or working odd hours. Supervisors and co-workers are taught to watch for warning signs, though privacy rules can complicate close monitoring.

## Safety with Emerging Technologies

As technology grows more advanced, new safety and security issues appear:

1. **Autonomous Weapons**: If a drone or ground robot can act without direct human control, a malfunction or hacking could lead to unintended harm. The line between an accident and a deliberate strike might blur if the system "chooses" a target incorrectly.
2. **Directed Energy and Lasers**: Such weapons might damage someone's eyesight or start fires if misused or poorly calibrated.
3. **Hypersonic Missiles**: Because they travel so fast, operators have little time to confirm a target or abort a launch if there is an error.
4. **Biotech and Genetic Tools**: Some research areas could be weaponized if safety protocols fail, leading to dangerous pathogens leaking from labs.

Regulations often lag behind these breakthroughs, leaving grey zones where safety standards and best practices are not fully defined. This demands caution from researchers and regulators alike.

## Managing Hazardous Materials

Many weapons contain hazardous materials, from explosive chemicals to radioactive components in nuclear arms. Handling these materials poses unique challenges:

- **Radiation Safety**: Nuclear warheads and their related fuel must be shielded. Crews need protective gear and training to avoid exposure. If nuclear components are poorly secured, the fallout from an accident could be devastating.
- **Chemical Agents**: Some militaries still store old stockpiles of chemical weapons (awaiting destruction under treaties). Even modern smoke grenades contain chemicals that irritate or

can burn. Proper ventilation, sealed containers, and strict procedures reduce accidental releases.
- **Explosive Powders**: Gunpowder and more advanced explosives can ignite from friction, heat, or static. Factories use anti-static flooring, special clothes for workers, and thorough cleaning to avoid dust buildup.
- **Environmental Spills**: In case of a spill, well-trained teams must contain the material quickly. Delays or mistakes can harm local communities and wildlife for years.

International guidelines for safe disposal also matter. Old shells, bombs, and toxic byproducts need careful destruction or recycling, not just dumping in landfills or oceans. Accidents from old munitions discovered decades later remain a problem in former war zones.

---

## Policing Civilian Markets

In countries where civilians can own firearms, authorities must keep track of millions of privately held weapons:

- **Background Checks and Permits**: Potential owners might need a license, a clean criminal record, and a safety course. Some places add "cooling-off" periods before a purchase can be finalized.
- **Registration**: Storing each gun's serial number in a central database helps trace stolen or used-in-crime weapons.
- **Limitations**: Banning high-capacity magazines or certain assault rifles is one approach. Others let owners have such items but require extra paperwork.
- **Safe Storage Laws**: Mandates that gun owners keep firearms locked and unloaded, separate from ammunition. Non-compliance can lead to fines or losing the right to own a weapon.

Despite these rules, illegal trafficking still happens. Criminal groups bypass official checks, turning to black markets or "ghost guns." Law enforcement tries to crack down, but the sheer size of some national gun pools complicates the job.

## International Cooperation on Safety

Many safety and security issues cross borders. Nations work together to share lessons and coordinate efforts:

1. **Standards and Best Practices**: Forums exist where militaries exchange tips on safe storage or transport. A country that faced a depot explosion might share findings to prevent repeats elsewhere.
2. **Combined Training**: Joint exercises let partner armies practice accident responses together, so they can act swiftly if a real event occurs.
3. **Border Security**: Neighbors might form special task forces to block smuggling routes, watch suspicious flights, or share intelligence on criminal networks.
4. **Global Databases**: Organizations like Interpol help track stolen firearms or wanted smugglers.
5. **Nuclear and Chemical Conventions**: International bodies verify that states handle dangerous items properly and destroy old stocks.

Strong cooperation raises overall safety, though not all countries participate equally. Distrust between certain nations can hinder the free exchange of security-related data.

## Training and Safety Culture

No matter how thorough the written procedures, a strong **safety culture** is crucial. This means everyone takes safety seriously, from generals and CEOs down to new recruits or factory hires. A good safety culture often includes:

- **Open Reporting**: Workers can report near-misses or concerns without fear of punishment. This helps fix issues early.
- **Regular Drills**: Fire drills, chemical spill practice, or weapon jam scenarios keep teams sharp.
- **Rewards for Caution**: Recognizing staff who follow rules carefully or suggest improvements can reinforce positive behavior.
- **Clear Communication**: Everyone knows who to call if they see a hazard. The chain of command for emergencies is widely understood.

Leaders set the tone. If managers skip steps or ignore small violations, others will follow that example. Frequent refreshers help staff remember procedures and adapt to new gear or updated rules.

---

## Disaster Response and Contingency Plans

Even with planning, accidents or threats can strike. Governments and firms build contingency plans:

1. **Emergency Teams**: Specialized units that handle bomb disposal, chemical leaks, or nuclear incidents. They travel quickly to trouble spots with gear and protective equipment.
2. **Evacuation Protocols**: If an incident risks civilian harm, local authorities must know how to evacuate or shelter communities.
3. **Medical Preparedness**: Hospitals near weapon sites might train for burn victims, chemical inhalation, or blast injuries. Adequate supplies of antidotes, bandages, and blood are vital.
4. **Public Communication**: In a crisis, rumors can spread panic. Officials should give clear, accurate updates about what happened, how to stay safe, and when the danger will pass.

Testing these plans with drills can reveal hidden flaws—like communication gaps or not having enough emergency vehicles. A quick, well-coordinated response can save lives and limit damage.

## Cyber Defense of Weapon Systems

Given the digital aspect of modern arms, cybersecurity is a major safety concern:

- **Firewalls and Encryption**: Strict measures to protect command networks from outside intrusion.
- **Frequent Updates**: Software must be patched regularly to fix vulnerabilities. Hackers often exploit systems that are left unupdated.
- **Air-Gapped Systems**: Some militaries keep critical control networks physically separate (no internet connection) to prevent remote hacking.
- **Red Team Testing**: Expert hackers (the "red team") are hired to simulate attacks, discovering weak points so they can be fixed.

Still, advanced hackers may find ways around such defenses, especially if insiders help or if third-party suppliers have insecure systems. That is why continuous vigilance matters.

## Balancing Security with Openness

Strict safety rules and heavy security sometimes conflict with the need for efficiency or public transparency. For instance:

- **Slowdowns in Production**: Thorough background checks on new workers, repeated inspections of cargo, and secure shipping can add delays that raise costs.

- **Limits on Information**: A company might keep design details secret to protect from espionage, but this secrecy makes outside oversight harder.
- **Cost vs. Benefit**: Some officials weigh how much to spend on advanced security features (like biometric locks or high-tech cameras). If the risk is viewed as low, they might opt for cheaper solutions, even if it raises the chance of theft.

Responsible leadership tries to balance these factors, ensuring safe operations without crippling normal workflows. The idea is to reduce dangers to a manageable level, not to create a fortress-like environment everywhere.

## Public Pressure and Reporting

When accidents happen or when stolen weapons lead to crimes, public anger can rise. People ask how such incidents were allowed and demand better controls. Media coverage can force companies and governments to improve safety measures. Whistleblowers who reveal hidden lapses can spark investigations. Non-governmental organizations may track data on lost or misused weapons, pushing for tighter laws or oversight. All of this can drive reforms—though such changes often happen after a serious event has grabbed headlines.

## Continuous Improvement

Safety and security in the weapon industry must evolve as threats change. A few ways this happens:

- **Learning from Incidents**: After a depot fire or a hijacked shipment, thorough reviews identify mistakes. Recommendations then feed into updated guidelines.

- **Technology Upgrades**: Cameras, biometric readers, or advanced ID badges become more affordable over time, enabling better monitoring.
- **Joint Exercises**: Militaries from different countries practice responding to a simulated accident. Observers share notes, leading to new best practices.
- **Staff Education**: Regular training modules keep employees aware of new threats, like social engineering or updated storage rules.

Organizations that treat safety as a never-ending journey often fare better than those that see it as a one-time checklist item.

---

# CHAPTER 18

### EFFECTS ON CONFLICTS AND WARS

Weapons hold the power to shape conflicts in deep and lasting ways. Throughout history, changes in weapon technology have altered battle outcomes, caused shifts in alliances, and even decided the fate of entire nations. Modern arms can make wars more devastating or, paradoxically, help some countries deter fights altogether. In this chapter, we explore how weapons affect the course of wars, the decisions leaders make, and the lives of everyday people caught in the middle. We also look at how advanced technology can transform battlefields, sometimes ending conflicts quickly but often at a high human cost.

## How Weapons Influence Battle Outcomes

Weapons are crucial tools in warfare. While strategy, leadership, and morale also matter, firepower plays a big part in who wins. Different weapons can:

1. **Increase Range**: From spears to rifles to missiles, each step in range evolution lets forces strike from farther away, often with less risk to themselves.
2. **Boost Lethality**: More powerful explosives or higher rates of fire can overwhelm defenders quickly.
3. **Improve Speed and Mobility**: Vehicles like tanks and armored cars let armies bypass enemy lines, taking advantage of weaknesses. Aircraft further expand this mobility.
4. **Enhance Precision**: Guided bombs and smart missiles reduce the need for mass bombing. They allow hits on specific targets—like a key bridge or command post—without spraying the entire area.

These advantages can shorten battles or cause an enemy to collapse faster. However, they do not always guarantee swift victory. Determined foes might adopt guerrilla tactics, hiding in cities or rough terrain where high-tech arms lose some edge. A mismatch in technology also does not always lead to total domination, as seen in many historical examples where smaller forces prevailed through local knowledge and unconventional strategies.

## Escalation and Arms Races

The more one side accumulates advanced weapons, the more others may feel they must catch up. This can spark an **arms race**, where countries keep developing new arsenals to outdo each other. Famous cases include the Cold War competition between the United States and the Soviet Union, which led to enormous stockpiles of nuclear, chemical, and biological weapons. Such escalation can:

- **Consume Resources**: Money spent on arms might be needed for education, health, or infrastructure.
- **Fuel Tension**: Rivals watch each other's military expansions nervously, sometimes leading to fear and aggression.
- **Risk Accidental Conflict**: In high-alert settings, a false alarm could trigger a war if leaders suspect they are under attack.

Even outside big powers, regional rivalries can lead to smaller-scale arms races. Neighboring nations might match each other's tank divisions or missile systems, raising the stakes whenever diplomatic relations sour.

## Deterrence and Preventing War

Paradoxically, having powerful weapons can also stop certain wars before they start. The logic is that an enemy will think twice if they know they will be countered with equal or greater force. The concept of **deterrence** relies on convincing potential foes that aggression would be too costly:

1. **Nuclear Deterrence**: Nations with nuclear weapons often say they keep them to prevent any large-scale attack, based on the idea of "Mutual Assured Destruction." If both sides can destroy each other, neither side dares to strike first.
2. **Conventional Deterrence**: Well-equipped militaries can discourage invasions. Tanks, planes, and well-trained troops make any aggression risky.
3. **Alliance Support**: Smaller countries might rely on alliances with stronger powers who pledge to defend them, effectively borrowing that strong power's deterrent.

However, deterrence is not foolproof. Mistaken beliefs, overconfidence, or miscommunication have historically led to conflicts. Some leaders might gamble that their opponent will not respond as harshly as feared, or they might underestimate the rival's capabilities.

## Shaping Warfare Strategies

As weapons evolve, so do the tactics and strategies of war. Generals plan how to use available arms in the best way:

- **Combined Arms**: Modern militaries mix infantry, armor, artillery, and aircraft to cover each other's weaknesses. Tanks break through defenses, soldiers follow to secure territory, and planes or helicopters provide support from above.
- **Precision Strikes**: Smart bombs and missiles allow hitting command centers, radars, or airfields early in a conflict. By targeting vital nodes, one side can disrupt the other's ability to coordinate.
- **Electronic Warfare**: Armies use jammers or cyber attacks to blind enemy radars, interrupt communications, or hack weapon systems. This can neutralize advanced gear without direct firefights.

- **Remote Engagement**: Drones and long-range missiles reduce risk to human pilots. Soldiers can stay behind the front lines, controlling unmanned vehicles that scout or attack.

At the same time, adversaries might adapt by hiding in civilian areas, digging tunnels, or dispersing their forces so precision weapons have trouble finding them all. This cat-and-mouse dynamic keeps changing warfare methods over time.

## Impact on Civilians and Infrastructure

While armies may claim advanced weapons reduce "collateral damage" (harm to non-combatants), war still hurts civilians in many ways:

1. **Urban Combat**: Fighting in cities leads to high civilian casualties and destroys homes. Even guided missiles can hit the wrong targets if intelligence is flawed.
2. **Explosive Remnants**: Unexploded bombs or landmines remain for years, endangering children and farmers. Regions like Southeast Asia and parts of Africa still deal with munitions left from past conflicts.
3. **Forced Displacement**: When bombs fall or artillery hits towns, people flee. Refugee crises strain neighboring countries and humanitarian groups.
4. **Attacks on Infrastructure**: Strikes on power plants, water facilities, or roads can cripple a society. Civilians lose access to clean water, electricity, or hospitals. Recovery can be slow and costly.

Humanitarian laws attempt to protect civilians, but in practice, some warring parties ignore these rules. Advanced weapons can inflict heavy damage quickly, leaving the survivors with long-lasting scars on health, environment, and economy.

## Psychological Effects

Weapons are not just physical tools; they also shape minds during a conflict:

- **Fear Factor**: Loud explosions, the sight of tanks, or knowledge that an enemy has powerful missiles can demoralize troops or terrorize populations.
- **Propaganda**: States may show off advanced weapons in parades or media to project strength and scare adversaries.
- **Uncertainty**: If an enemy might have lethal drones overhead or snipers with long-range rifles, soldiers become jumpy, sometimes leading to panicked firing or stress disorders.
- **Resistance and Unity**: On the other hand, seeing a powerful enemy can unite local populations to resist together, fueling a sense of defiance.

Leaders often plan how to use the psychological impact of weapons. A single demonstration shot or a small-scale strike might break an opponent's will to fight, avoiding a larger battle. However, it can also harden resentment if innocent people are harmed.

---

## Lengthening or Shortening Wars

Better weapons do not always end wars quickly. Sometimes they make it easier for a weaker side to continue fighting:

1. **Asymmetric Conflicts**: Guerrillas or insurgents use portable weapons—like shoulder-fired missiles or improvised explosives—to ambush a stronger army that has jets and tanks. This can drag out the fighting for years.
2. **Proxies and Supply Lines**: External powers might provide arms to local groups, enabling them to keep resisting. Such proxy wars can linger, as new shipments keep arriving.

3. **Fortified Defenses**: If defenders build bunkers, tunnels, and anti-air systems, the attacker may face a long stalemate despite advanced gear. Siege tactics, blockades, and repeated bombings can follow.

On the flip side, some high-tech arms can achieve rapid success—like sudden missile barrages that destroy key facilities. Once an opponent's command structure or supply routes collapse, the war might end sooner. The deciding factor is often how well each side adapts and how deep their resources run.

## Weapons of Mass Destruction

Weapons of mass destruction—nuclear, biological, or chemical—raise the stakes far beyond conventional arms:

- **Nuclear War**: Even a limited exchange can kill millions and affect the global climate. Fear of these catastrophic effects has prevented their regular use since World War II, but the risk remains.
- **Chemical Agents**: Poison gases and toxins can spread without damaging buildings, but cause horrific injuries. Worldwide treaties ban them, yet some regimes or terror groups may still use them.
- **Biological Threats**: Infectious pathogens designed to spread among populations can be harder to contain. They are also unpredictable; once a virus escapes a lab, it might not stop at enemy lines.

These weapons make many conflicts extremely dangerous, possibly turning local fights into disasters for entire regions or the planet. Thus, global efforts aim to control or eliminate them—though not all parties cooperate.

## Role of Arms Exports and Aid

Countries often send weapons to allies involved in conflicts:

1. **Direct Aid**: Free or discounted arms shipments help a favored side fight. This can tip the balance in civil wars or regional disputes.
2. **Loan or Lease**: Nations might lend tanks or planes, with the recipient paying over time or returning them later.
3. **Training and Advisors**: Alongside weapons, the supplier sends military experts to show how to use the equipment effectively.
4. **Covert Support**: Governments might deny involvement but secretly ship arms to certain groups, hoping to influence the outcome or create problems for a rival.

Such interventions sometimes widen conflicts, drawing in more players. If the receiving side commits abuses, the supplier risks a public relations crisis. Even after wars end, leftover weapons can fuel organized crime or new rebellions.

---

## Changing Face of Battlefields

Modern warfare can look very different from classic trench battles:

- **Drone Warfare**: Reconnaissance drones gather intel without risking pilots. Armed drones strike with precision, though critics say they can cause civilian casualties if intelligence is flawed.
- **Cyber Battles**: Hackers aim to cripple radars, disrupt power grids, or confuse enemy communication. A war might begin with a wave of cyberattacks before any physical attack.

- **Space-Based Assets**: Satellites track enemy movements, provide GPS data, and relay commands. Destroying or jamming satellites can blind an opponent.
- **Hybrid Tactics**: Adversaries mix conventional fights with propaganda, sabotage, and economic pressure. Weapons are only one piece of a multi-layered conflict.

This shift does not mean classic weapons vanish. Rifles, mortars, and simple artillery remain widespread, especially in poorer regions. It does mean advanced militaries use a greater variety of tools to gain an edge.

## Civil Wars and Insurgencies

Civil wars often see a patchwork of militias, government forces, and foreign sponsors. Weapons flow in from many sources:

1. **Rebel Weapon Seizures**: Insurgents capture gear from government depots or battlefields.
2. **Black Market**: Smugglers bring guns from abroad, sometimes financed by illegal trade in drugs, diamonds, or other goods.
3. **External Backers**: A foreign power might give arms to a rebel group that shares its goals.
4. **Improvised Devices**: Rebels with limited resources build homemade bombs or adapt civilian drones for attacks.

This patchwork can lead to messy, prolonged conflicts, as no single side can easily outgun the others. Civilians suffer greatly, and even once peace deals come, disarming groups may prove tough. Piles of small arms remain scattered throughout society.

## Peacekeeping and Arms Control

In some conflicts, international peacekeepers step in, attempting to separate warring sides and reduce violence. Their effectiveness often relies on:

- **Rules of Engagement**: Whether peacekeepers can use force to disarm militias or if they only observe. A robust mandate can help them seize illegal arms.
- **Weapon Collection**: Disarmament programs might offer money for turning in guns, or supervise the destruction of heavy weapons.
- **Confidence-Building Measures**: Agreements on reducing troop numbers, pulling back artillery, or verifying compliance with monitors.
- **Long-Term Solutions**: Addressing the root causes of conflict—like economic grievances or political exclusion—so that people do not rearm later.

Even strong peacekeeping missions can struggle if local factions do not genuinely want peace or if big powers keep sending new arms into the region.

## Humanitarian Law and War Crimes

International law tries to limit the worst effects of war. The Geneva Conventions and other treaties ban targeting civilians on purpose, or using weapons that cause unnecessary suffering (like certain chemical agents). Violations can lead to charges of **war crimes**. Even if enforcement is patchy, fear of future prosecution may restrain some leaders or officers. Advanced surveillance and forensic methods (including satellites and digital tracking) make it harder to hide large-scale abuses. However, these laws rarely stop a determined actor if they believe they will not be caught or punished.

## Technological Gap and Global Inequality

Wealthy nations can afford cutting-edge jets, drones, and missile defenses. Poorer countries might rely on old Soviet-era tanks or cheap rifles. This gap has consequences:

- **One-Sided Fights**: If a strong power intervenes against a weaker adversary, the latter has few conventional ways to respond. They may resort to terrorism or guerrilla warfare.
- **Dependence on Foreign Suppliers**: Nations without a local arms industry must buy from abroad, which can trap them in debt or lead to political pressure from the supplier.
- **Brain Drain**: Talented engineers in poorer regions might migrate to places with advanced defense industries, leaving fewer skills at home.

Some countries try to leap ahead by developing niche technologies (like ballistic missiles or cyber warfare) to offset the imbalance. This can create new security dilemmas if neighbors feel threatened.

---

## Long-Term Effects After a Conflict

Even when wars end, the presence of weapons can shape the aftermath:

1. **Post-War Militarization**: Large armies may remain. Veteran fighters keep their guns or adapt them for criminal activities. Police might be outgunned.
2. **Slow Demobilization**: Governments must collect, store, or destroy surplus arms. Failure to do so lets arms slip into black markets.
3. **Economic Recovery**: Destroyed infrastructure and farmland make rebuilding hard. War veterans need jobs or rehabilitation.
4. **Social Trauma**: Civilians who lived under bombing or mortar attacks may deal with mental health issues for years. Families displaced by violence can struggle to resettle.

5. **Political Resentment**: If a foreign power used advanced weapons that caused widespread destruction, local people might hold long-lasting anger, fueling future radicalization.

Addressing these problems requires disarmament, rebuilding, and reconciliation efforts—a process that can take decades.

## Case Studies in Rapid Change

Some historical conflicts show how a single new weapon altered a balance:

- **Machine Guns in World War I**: Made mass infantry charges deadly, forcing trench warfare.
- **Atomic Bomb in World War II**: Ended the war with Japan swiftly but opened the nuclear age, reshaping global politics.
- **Stinger Missiles in the Soviet-Afghan War**: Rebels shot down Soviet helicopters, shifting momentum.
- **Drones in Recent Wars**: Let advanced armies perform strikes without risking pilots, but also caused local backlash over civilian deaths.

Each case underlines how technology can disrupt existing battle plans and lead to new warfighting methods.

## Ethics of Advanced Weapons

Modern arms like autonomous drones or cyber tools raise ethical questions:

- **Should Machines Decide?** If an autonomous weapon picks targets on its own, is that moral? Who is accountable if it attacks civilians by mistake?

- **Distance and Responsibility**: Pilots firing missiles from a drone control room thousands of miles away might feel detached from the reality of killing. Does this encourage riskier strikes?
- **Unseen Collateral Damage**: Cyber attacks might knock out hospitals or water systems, harming civilians indirectly. Is that acceptable under international law?
- **Arms for Oppressive Regimes**: Companies selling to governments that abuse human rights face moral scrutiny, but often the deals go ahead due to political or economic interests.

As weapons become more sophisticated, these debates intensify. Some want bans on "killer robots" or stricter treaties for cyber warfare, while others argue technology itself is neutral, and only the user's intentions matter.

## Efforts to Limit Destructive Wars

Faced with the destructive potential of modern arms, nations and organizations attempt various solutions:

1. **Arms Control Treaties**: Agreements to reduce or ban certain categories, like nuclear or chemical weapons.
2. **Inspections and Verification**: Checking compliance, often via international teams.
3. **Dialogue and Diplomacy**: Summits where rival states negotiate mutual limits on forces or confidence-building measures.
4. **Peacekeeping**: International bodies may place troops as buffers.
5. **Conflict Resolution Programs**: Mediation and negotiations to address underlying causes, hopefully reducing reliance on brute force.

The success of these efforts depends on political will. If states see more advantage in building arms than limiting them, treaties can collapse. Yet history shows that sometimes warring parties do choose restraint when they realize the catastrophic cost of modern warfare.

## Conclusion of Chapter 18

Weapons undeniably shape conflicts and wars. They can bring quick wins or drag fights into long stalemates, cause untold civilian suffering, or scare enemies into backing down. Their influence goes beyond the battlefield, affecting economies, global power balances, and the day-to-day reality of people living in war zones. The arms race cycle persists: countries fear falling behind in technology, so they invest heavily, which in turn spurs others to do the same.

Some argue that advanced weapons keep peace through deterrence, preventing major wars between powerful nations. Others point out the countless local or civil conflicts worsened by easy access to lethal arms. The arrival of new technologies—cyber attacks, AI-driven drones, or hypersonic missiles—suggests that future warfare could become even more complex and unpredictable.

Ultimately, how weapons are used depends on human choices. Leaders, soldiers, and the societies that build and fund militaries play pivotal roles in deciding whether advanced arms serve to defend against threats or to start destructive campaigns. As long as conflicts arise from political, social, or economic tensions, weapons will continue to play a key role—both as a factor of devastation and, in some cases, as a deterrent that keeps hostilities at bay. Understanding these dynamics is vital for anyone trying to reduce suffering and steer humanity toward resolving disputes without constant violence.

# CHAPTER 19

## QUESTIONS ABOUT ETHICS AND MORALS

Weapons hold great power. They can protect whole communities or cause incredible destruction. How they are used, who controls them, and why they exist lead to many questions about right and wrong. These moral and ethical concerns do not have simple answers, because people hold different values, beliefs, and perspectives. In this chapter, we will look at the wide range of moral issues that surround weapons and the industry that makes them. We will explore concerns about human suffering, responsibility, fairness, and the deep worries that modern arms—especially advanced or autonomous ones—raise for our global community.

## 1. The Value of Life and the Need for Defense

One of the first moral questions about weapons is why they exist at all. Most people agree that life is precious. Yet, many also feel that having the ability to fight is sometimes necessary, whether it is to fend off aggressors, prevent abuses, or stand up for principles like freedom.

- **Right to Self-Defense**
  At the heart of many moral discussions is the idea that individuals, groups, or nations have the right to defend themselves if threatened. This justifies owning weapons in certain laws or traditions. Some believe it is a basic right: if someone tries to harm you, you can fight back.
- **Use of Force**
  Even if defense is justified, there is debate about what level of force is moral. Should one use lethal weapons against

thieves? Is it ever right to bomb an enemy city and risk civilian lives? Does owning a gun at home for protection outweigh the risks of accidents or misuse? Answers vary across cultures and legal systems.

- **Balancing Safety and Harm**
  People who argue for stronger defense capabilities say that robust military forces help maintain peace by scaring off potential attackers. Critics counter that stockpiling arms can make it easier for leaders to choose violence and might increase the risk of dangerous misunderstandings.

In each of these areas, moral arguments often rely on values: the protection of innocent people, the prevention of oppressive rule, or the belief that peaceful solutions should be found before turning to force. There is no single moral code that everyone agrees on, which is why weapons remain such a heated topic in every society.

---

## 2. Weapons and Human Suffering

Moral debates also center on how weapons cause suffering during and after conflicts. History is filled with examples of towns destroyed, families displaced, and long-lasting trauma.

- **Civilian Casualties**
  Even with "smart" bombs and careful targeting, civilian deaths happen. Some argue that this is unavoidable in war, but many question whether these deaths can be morally justified. Is it right to bomb a city if there is a chance that children might be present? Do we weigh the risk to our own soldiers more heavily than the risk to enemy civilians?
- **Long-Term Wounds**
  Even if a war ends, landmines can remain in fields for decades, injuring farmers or children. Chemical or nuclear

fallout can poison environments, harming health and food supplies. The moral dilemma is not just about the immediate use of weapons but about their lingering effects on people who had no say in the conflict.

- **Torture and Cruel Methods**
  Some weapons are judged inhumane because they cause extreme pain or long-lasting harm. For instance, devices that create mass burning or chemical injuries are often condemned by international treaties. Critics wonder if any war method that leads to severe suffering should be banned, or if all warfare is inherently cruel.

These points lead many ethicists to argue for strict limits on how weapons are used, or for banning certain types altogether. The more a weapon indiscriminately harms civilians and leaves behind chaos, the stronger the moral objection tends to be.

---

## 3. Arms Production and Profit

An additional moral question is whether it is right to make money from selling tools of violence. Defense companies can earn large profits by selling to governments or private buyers. While this is legal in many places, critics question the morality of profiting from products that can kill.

- **Job Creation vs. Harm**
  Advocates say the weapon industry provides jobs and technological development. Workers in these companies are able to support families, and local economies may flourish. Yet, the final product could be used in ways that bring devastation elsewhere. Should the benefit to some communities justify the potential harm to others?

- **Shareholders and Responsibility**
  Some investors avoid putting their money into arms manufacturers, feeling that it conflicts with ethical principles. Others see no problem if the buyer is a legitimate government. But what if that government has a record of abusing human rights? Does the company share blame if those weapons are later used against peaceful protesters?
- **Human Rights Concerns**
  Many controversies arise when weapons are sold to regimes known for oppression or conflicts with high civilian casualties. Even if the deal is legal, is it moral to deliver advanced arms to a leader who might turn them on innocent people? The debate grows louder if the selling country claims to stand for human rights.

These questions place the arms trade under constant scrutiny. Companies insist they follow all laws and have no control over foreign policies, but human rights groups often say that just because something is lawful does not make it moral.

---

## 4. Responsibility of Scientists and Engineers

People who design or build weapons might struggle with their personal moral compasses. They have special knowledge and skills, which can be applied to medicine or communication devices instead.

- **Moral Burdens**
  Should an engineer refuse to work on a missile guidance system if they believe it might be used to hurt civilians? Some do, citing moral objections or personal beliefs. Others argue that someone else will do it if they do not, so they might as well do the best job possible.

- **Dual-Use Research**
  Scientists might discover new materials or methods useful in many fields, from medical equipment to bomb-making. They could fear their discoveries will be weaponized. The moral weight is heavy: do they stop researching or keep going, hoping the good uses outweigh the bad?
- **Codes of Conduct**
  Some professional societies encourage codes of ethics discouraging members from contributing to harmful weapons. But these guidelines are not always enforced. Countries vary in how they see scientific freedom, national defense needs, and personal conscience.

Thus, individuals working in the weapon sector face moral tensions between their careers, their loyalty to their country, and broader concerns about global well-being.

---

## 5. Nuclear, Biological, and Chemical Weapons

Nuclear bombs, deadly viruses, and toxic agents can kill vast numbers of people, sometimes without distinguishing between soldiers and civilians. Their existence raises ethical alarms louder than almost any other topic in the arms world.

- **Mutual Assured Destruction**
  Countries with nuclear weapons often argue they keep them only to deter attacks, not to use them. But accidents, miscommunication, or irrational decisions could unleash catastrophic results. Many see no moral justification for risking the survival of entire regions—or the planet.
- **Treaties and Bans**
  International agreements aim to ban or limit these weapons because the harm they cause is seen as too great. But not all

nations sign these treaties, and enforcement is inconsistent. Some regimes see them as the ultimate security guarantee. Ethicists ask: Is it ever moral to threaten total devastation to keep peace?

- **Spreading Risk**
  Biological weapons could mutate or spread beyond control. A virus unleashed for warfare does not obey borders, potentially sparking a global crisis. Chemical weapons can poison land, water, or air. The moral question is whether any short-term military advantage can ever justify the uncontrollable damage these tools might cause.

Overall, the presence of weapons of mass destruction puts the entire moral debate on a different scale, turning local or regional conflicts into global risks. Their severe dangers lead many to call for total elimination, yet nations hesitate to give up what they see as their strongest deterrent.

## 6. Autonomy and "Killer Robots"

As technology advances, some weapons can function with minimal human involvement. Drones already fly missions guided by remote operators, and research is pushing toward fully autonomous systems that decide on targets themselves.

- **Removing Humans from the Loop**
  Advocates say robot soldiers or drones reduce risk to one's own forces and might make faster, more accurate decisions. But if a machine malfunctions or wrongly identifies a target, who is morally liable? A software flaw could lead to an unplanned attack, harming civilians.
- **Accountability**
  War crimes typically place blame on commanding officers or politicians. With autonomous weapons, does blame lie with

the programmer, the company that made the AI, or the operator who launched it? Many fear a gap in accountability, as each party can blame another.
- **Ethical Campaigns**
Organizations across the world call for a ban on "killer robots," urging that humans must always make life-and-death decisions, not algorithms. Others say well-designed AI might be less prone to panic or error than humans, thus saving lives overall. This debate highlights the moral complexity that arises when technology outstrips the rules we have in place.

## 7. Arms Trade to Unstable Regions

Large-scale sales of weapons to areas suffering from civil war or poverty raise stark moral questions. If guns flood a region that lacks strong institutions or has corrupt leadership, the results can be tragic.

- **Profit vs. Suffering**
Are arms dealers morally responsible if a buyer misuses the weapons? Some argue yes, especially if there were clear signs of likely abuse. Others say the seller cannot control how a buyer acts.
- **International Rules**
Treaties like the Arms Trade Treaty aim to stop sales that might escalate violence or violate human rights. Yet, big defense exporters sometimes find ways around these rules by issuing exceptions or denying knowledge of potential abuses.
- **Humanitarian Disasters**
In the worst cases, arms delivered to an unstable government are used to crush dissent or fuel cross-border invasions. This leads to refugees, famine, and regional instability. Do the countries or companies that sold the arms have a moral duty to help address the crisis they contributed to?

Such tragedies can sway public opinion against arms exports, prompting governments to adopt stricter checks. But as soon as a new threat emerges, calls for stronger allied militaries can push those checks aside.

## 8. Culture, Religion, and Moral Beliefs

In many cultures, there are deep-seated views on violence and conflict. Some religious traditions strongly oppose any use of weapons, promoting non-violence as a sacred principle. Others allow force if it meets certain conditions, like self-defense or protecting innocent lives.

- **Pacifism**
  Some groups believe all violence is morally wrong. They see weapons as symbols of a broken human condition that must be overcome through peace, negotiation, or faith in non-violent protest. Pacifists sometimes refuse military service or employment in arms factories.
- **Just War Theory**
  Within certain religious or philosophical frameworks, war can be considered just if it meets strict criteria: it must be declared by a legitimate authority, have a just cause (like defending against aggression), and use proportional force that spares civilians. Under these rules, weapons are morally acceptable, but only up to a point.
- **Secular Moral Codes**
  Even outside religion, ethical philosophies debate the morality of violence. Some believe that if a war prevents a greater evil (for instance, stopping genocide), using weapons is morally justified. Others hold that the cycle of violence only grows when arms are introduced, making peaceful methods the only moral choice.

Cultural differences also show up in gun ownership among civilians. In some places, owning firearms is seen as a right linked to freedom and self-reliance. In others, the idea of widespread private gun possession is viewed as dangerous and irresponsible. These beliefs shape laws and moral norms.

## 9. Balancing Individual and Collective Morality

Some moral questions revolve around individuals—like a soldier deciding whether to follow an order they find unjust. Others center on entire nations or companies, such as whether a government should sell weapons to an ally accused of war crimes. Balancing personal and collective morality can be hard:

1. **Following Orders vs. Conscience**
   Soldiers are trained to obey commands, but what if they believe a mission is unethical? Some armies allow conscientious objection, but in many cases, refusing orders can lead to punishment.
2. **Patriotism vs. Global Responsibility**
   Citizens may feel proud if their country's defense industry succeeds, providing jobs and boosting economy. Yet the same weapons could be used in ways that clash with broader human rights.
3. **Corporate Goals vs. Ethical Stands**
   A defense firm's board might see a lucrative contract. Some staff, though, might raise moral objections, especially if the buyer has a poor human rights record. Should the company turn down profits for moral reasons?
4. **Consumers and Voters**
   Even everyday people might have a role. If citizens vote for leaders who promise to cut arms exports to abusive regimes, or if consumers boycott certain brands, moral pressure can shift corporate or government behaviors.

These tensions show that ethics in the arms world often cross from personal conviction into major policy decisions. Collective action or inaction can make the difference between forging a path that tries to reduce harm and one that tolerates it for perceived gains.

## 10. Searching for Ethical Standards

Across the globe, individuals and organizations work to define or enforce moral standards for weapons:

- **International Treaties and Conventions**
  While these documents are legal, they also reflect shared moral beliefs that certain behaviors—like torture or targeting hospitals—are unacceptable. They create frameworks for accountability.
- **Global Grassroots Movements**
  Peace groups, religious bodies, and ethical investors push for disarmament or stricter controls. They hold protests, meet lawmakers, and run awareness campaigns. Their moral arguments try to influence public opinion and policy.
- **Corporate Ethics Programs**
  Some big arms companies adopt rules for responsible sales. They might pledge not to deal with parties under UN embargo or those known for atrocities. Skeptics question if these pledges truly limit business or are just public relations moves.
- **Academic and Policy Debate**
  Scholars in ethics, law, and political science publish articles exploring topics like artificial intelligence in warfare, the moral cost of nuclear deterrence, and the duty to protect civilians. Governments sometimes rely on these studies when revising policies.

While none of these efforts fully solve the moral challenges, they shape how societies talk about weapons. They remind governments and businesses that behind every rifle or missile are deep moral stakes—real people's lives, families, and futures.

## 11. Critiques of Moral Debates

Some voices argue that focusing on ethics in the weapon industry is pointless, because war itself is about power, not morality. They claim that attempts to inject moral principles into a fundamentally violent field do not stop conflicts. Others say the moral approach is naive:

- **Realist Perspective**
  In international relations, "realists" see states as acting mainly on self-interest. They believe moral or ethical talk is overshadowed by the pursuit of security or advantage. If it helps them, states will use or sell weapons, no matter what.
- **Strategic Necessity**
  Military planners might see morality as less important than winning. If the choice is losing or using lethal force, they pick force. They might see regrets as unfortunate but necessary.
- **Practical Limits**
  Even if a society wants to uphold high moral standards, enemies might exploit that kindness. For example, if one side avoids bombing certain sites, but the other side hides soldiers there, who is morally right? The conflict's messy nature complicates applying absolute moral rules.

Such critiques highlight how moral debates, while important, do not always translate smoothly into policy or battlefield decisions. Nonetheless, many people and governments still believe moral guidelines can shape better outcomes or at least reduce atrocities.

## 12. Future Ethical Dilemmas

As weapon technology evolves, moral questions will keep growing:

1. **Gene-Editing for Soldiers**: Could future armies use genetic engineering to create soldiers with greater endurance or fewer feelings of fear? Is that ethically right?
2. **AI Leadership**: If an algorithm plans entire campaigns, picking targets and scheduling attacks, is it moral to remove human judgment from warfare strategy?
3. **Space Weaponization**: Some fear the day lasers or kinetic weapons in space might target Earth or satellites. Does cosmic defense cross ethical lines in ways we have not yet prepared for?
4. **Climate War**: Rising temperatures and resource shortages might fuel conflicts. Who is morally at fault if they use advanced arms to secure water or farmland?

These scenarios reveal how moral discussions never end; they adapt to new possibilities. Proactive regulations, combined with global dialogue, might guide technology away from misuse. But the lure of power or strategic advantage can overshadow these moral concerns, leaving the future uncertain.

---

# CHAPTER 20

## POSSIBLE FUTURES OF THE WEAPON INDUSTRY

Looking into the future of the weapon industry means considering technology, politics, economics, and social forces, all interacting in complex ways. While we cannot predict exactly what will happen, we can analyze trends and imagine where things might lead. Some paths could bring stricter regulation, new waves of disarmament, and advanced protective measures. Others could see an arms race in more areas—such as space, artificial intelligence, or genetic enhancement.

In this final chapter, we look at the key factors that will shape the weapon industry's future. We cover how changing world powers, emerging technology, environmental pressures, and shifting public attitudes might direct the course of research, production, and global weapons trading. Through these topics, we see that the industry's future is not fixed but influenced by many human choices—choices that can expand or restrict the role of force in our global community.

## 1. Continuation of High-Tech Arms Races

One possible future is that competition among major powers will remain high. As nations race to lead in artificial intelligence, quantum computing, and space technology, they could pour money into advanced weapon development.

- **Quantum and AI Breakthroughs**
  If a country develops quantum-based sensors or communication tools, it might locate submarines or stealth aircraft more easily. AI could manage battlefields in real time,

analyzing data too vast for human planners. This might spark a new form of arms race where each side strives for the smartest, fastest systems.

- **Hypersonic Missiles**
  Missiles that travel multiple times the speed of sound are already in early deployment. They reduce reaction time for defense systems, potentially increasing the risk of miscalculations. If multiple countries widely adopt them, we could see new defense technologies to counter them, raising costs and tensions.

- **Cyber and Information Warfare**
  Future conflicts might focus on disabling an opponent's networks or sowing confusion rather than launching large-scale bombing campaigns. Weapons may evolve to become digital: malicious code that can sabotage power grids, factories, or even advanced weapon control systems.

In such a scenario, the industry would likely become more secretive, with major powers guarding their tech. Traditional tank or jet makers might partner with software and data analytics companies. Military spending could remain high, driven by fear that falling behind spells vulnerability.

---

## 2. Growth of Private Military and Security Markets

Private security companies, some armed with sophisticated tools, already operate worldwide. In the future, these firms could expand:

- **Corporate Armies**
  Large corporations might hire private military contractors to protect their assets—like mines or refineries—in unstable regions. If governments struggle to maintain order, businesses may take security into their own hands.

- **Rogue Element Risk**
  More armed private groups means a higher chance that weapons or trained personnel drift toward illegal actions or become mercenaries for hire. This might spread chaos, especially if states allow it for quick solutions.
- **Global Standards or Lack of Them**
  Efforts to regulate private security across borders might emerge, but enforcement is tricky. An international code of conduct could set rules, but not every company or client would obey. The weapon industry might focus on making gear specifically for these private groups, from small drones to advanced surveillance kits.

If private security grows unchecked, some worry it erodes the state's monopoly on force, undermining democracy and accountability. Others see it as a response to the state's failure to keep people safe. Either way, the weapon industry might find a booming market in this direction.

---

## 3. Emphasis on Less-Lethal and Non-Lethal Tools

Another trend could be an increased push for weapons that incapacitate targets without killing them. Police forces, security services, or peacekeeping missions might want advanced tools that reduce fatal outcomes:

- **Directed Energy Devices**
  Research on lasers or microwave beams that disable electronics or cause temporary pain could grow. These might be used for crowd control or to stop vehicles, aiming to avoid lethal force.
- **Sticky Foam, Net Guns, and Drones**
  Non-traditional gear that immobilizes suspects or vehicles might become more common. For example, a net launched by

a drone to entangle a fleeing car's wheels or a foam that hardens around a criminal's limbs.
- **Chemical or Acoustic Devices**
Some systems deliver irritating chemicals (like pepper spray) or disorienting sound waves. The moral question is whether they truly minimize harm or cause unintended injuries or permanent damage to hearing or lungs.

While less-lethal products have been around for a while, advanced technology might make them more reliable. Some police and military units prefer these options to reduce civilian casualties and negative publicity. The weapon industry, seeing an opportunity, could expand this sector. Still, critics point out that even "less-lethal" arms can be misused, and the line between safe and harmful can be blurry.

---

## 4. Rising Importance of Space and Orbital Assets

With satellites guiding missiles, supporting communications, and gathering intelligence, space is already linked to modern warfare. In the future, protecting or targeting orbital assets could define new battlegrounds:

- **Satellite Defense**
Countries might invest in anti-satellite weapons to blind an opponent's observation or signals. This includes ground-based missiles, laser systems, or even small "kamikaze" satellites that collide with enemy satellites.
- **Space as a Frontier**
Some worry that as humans plan to colonize or mine the Moon or asteroids, militaries will follow to claim resources or strategic positions. Weapons could be placed on lunar outposts or in orbit.

- **Global Tensions**
  If certain powers threaten to knock out rivals' satellites, conflicts might escalate quickly. A strike in space can disrupt global communications, harming civilians as well. Companies might develop new lines of space-based surveillance or weapons, adding complexity to arms control.

Efforts to keep space free of weapons have existed for decades, but with limited success. If the arms industry sees financial gains in space-based systems, lobbying might push for looser restrictions, opening a whole new domain of conflict.

---

## 5. Eco-Driven Conflicts and Environmental Pressures

Climate change can bring water shortages, extreme weather, and food crises, which might spark wars or mass migration. The weapon industry would likely adapt to these shifting conflict conditions:

- **Resource Wars**
  If countries fight over scarce water, farmland, or mineral deposits, they may invest heavily in specialized equipment for harsh conditions—like drones able to operate in extreme heat or amphibious vehicles that navigate flooded areas.
- **Disaster Response**
  Some armies see themselves as first responders to natural disasters. The weapon industry might produce dual-purpose vehicles that can deliver aid or switch to defense roles quickly.
- **Green Weapon Production**
  Environmental concerns might also shape how arms are made, focusing on reducing pollution from factories or designing less polluting equipment. This might be superficial or a genuine shift, depending on market pressure and regulations.

This future merges security and sustainability, with the industry providing tools for both. But moral questions arise if the environment is weaponized—like using a dam breach to flood an enemy region. The ethics of such acts would be heavily debated in the global arena.

## 6. Stronger Global Controls and Disarmament Movements

It is also possible that the global community takes a firmer stand against the spread of dangerous arms, creating tighter rules and seeing more countries join disarmament programs:

- **Renewed Anti-Nuclear Efforts**
  If nuclear weapons are deemed too risky in a multipolar world, major powers might sign new treaties to reduce arsenals further, with strict verification. This would slow the nuclear arms industry, though it might still thrive for civilian nuclear technology.
- **Ban on Autonomous Killing Machines**
  Public outcry or fear of AI going rogue might lead to an international treaty banning fully autonomous lethal weapons. Such a treaty could drive companies to focus on systems that keep a human in control.
- **Crackdowns on Illicit Arms Trade**
  Better technology (like blockchain shipping records or satellite surveillance) could catch smugglers more easily. More robust enforcement by the UN or regional alliances might reduce black-market flows, though criminals would still look for loopholes.
- **Peace Dividend**
  If global tensions ease in certain regions, governments might cut defense budgets, redirecting funds to social programs. This would reduce arms sales. Some defense firms might

adapt by focusing on civilian products or high-tech fields like robotics without military uses.

While disarmament movements have faced setbacks in the past, changing public opinion or major crises—like near-nuclear war scares—can revive them. Still, achieving strong global controls requires broad trust and cooperation, which is often in short supply.

## 7. Small Arms Proliferation and Civil Markets

On the civilian side, small arms could still cause turmoil in many places:

- **Technological Advances in Firearms**
  Smart guns that only fire for authorized users might become more common. Some states might mandate these to reduce theft or misuse. Others resist such rules, seeing them as infringement on personal freedoms.
- **3D-Printed Weapons**
  As 3D printing gets cheaper and better, anyone with the right designs could print components for firearms. This could blur the lines between legal and illegal production, making regulation harder.
- **Global Criminal Networks**
  Demand for small arms by gangs or insurgents might stay high, fueling smuggling routes. Governments try to clamp down with data-sharing and advanced detection technologies, but the black market is adaptable.
- **Private Gun Ownership Debates**
  In some countries, there could be tighter rules following public tragedies. In others, political traditions might keep ownership broad. The weapon industry might shift to producing accessories, non-lethal options, or advanced lock systems as laws evolve.

Thus, small arms remain a core part of the industry's future, whether for personal defense, law enforcement, or criminal use. Governments that want more control face pushback from lobby groups and cultural values favoring personal rights.

## 8. Ethical and Public Pressure

As awareness grows about how weapons can be misused, the public might demand higher standards from arms producers and their governments:

1. **Shareholder Activism**
   Investors could demand transparency or stop funding companies that sell to regimes involved in war crimes. Firms might respond by screening clients more carefully.
2. **Global Campaigns**
   Peace groups and social movements may target big arms fairs, protest outside manufacturing plants, or boycott corporations involved in controversial sales. Media coverage intensifies scrutiny.
3. **Legal Cases**
   Victims of bombings might sue weapon makers if they believe the firm sold arms recklessly. Courts could set precedents that require stricter due diligence.
4. **Consumer Technology Impact**
   If large consumer tech companies are also partners in military AI or robotics, they might face employee walkouts or public criticism. Some might choose to quit certain defense contracts.

This future suggests that moral or social accountability will not just be theoretical. It might shape the industry's bottom line, pushing it to adopt self-imposed rules, or possibly leading to big shifts in how weapons are marketed and sold.

# 9. Reshaping Military Doctrine

As technology changes or if strong disarmament measures arise, militaries might transform their doctrines and structure:

- **Lean, High-Tech Forces**
  Nations could opt for smaller but more advanced armies, relying on drones, robots, and cyber teams rather than large numbers of infantry. This approach might cut manpower costs but raise R&D spending.
- **Multi-Domain Operations**
  The concept of fighting across land, air, sea, space, and cyberspace simultaneously is growing. Future militaries may train in integrated battle scenarios where controlling data flows is as vital as controlling territory.
- **Deterrence by Denial**
  Some states might develop layered defense systems—like anti-drone fences, railgun interceptors, or energy shields—to block attacks rather than retaliate. This aims to make aggression futile, though the technology is still in early stages.
- **Civil-Military Fusion**
  More advanced weapons could come from civilian tech labs, with direct ties to the armed forces. This might hasten innovation but raise questions about mixing commercial culture with lethal aims.

The weapon industry will respond to these doctrinal shifts by creating products to fit new military needs—smart helmets with augmented reality, quantum encryption devices, or stealthy supply drones for tricky terrain.

## 10. Humanitarian and Environmental Focus

A hopeful possibility is that societies demand major changes that reduce warfare and emphasize global well-being:

1. **Conflict Resolution**
   Diplomatic methods, economic partnerships, and cultural exchange could lower tensions, diminishing the arms market. Large militaries might see less justification, especially if countries find cooperation more profitable than confrontation.
2. **Defense Diversification**
   Weapon makers could pivot to peaceful technologies. For instance, the same engineering that builds fighter jets might adapt to eco-friendly aircraft or advanced medical devices. This process, sometimes called "conversion," might require policy incentives and strong political will.
3. **Climate Collaboration**
   If climate disasters force humanity to collaborate, major powers might reduce arms spending to invest in shared solutions for energy, water, and disaster prevention. The weapon industry might shrink in the face of shifting priorities.
4. **Global Policing**
   Alternatively, there could be a large global force under the UN or a similar body, aiming to keep peace worldwide. The arms industry could supply equipment for this force, regulated under strict guidelines that ban certain destructive tools.

While these visions may sound optimistic, history shows that major changes can happen when public opinion and leadership align—especially after crises that reveal the high cost of ignoring global threats.

## 11. Potential Game-Changers

Some single events or breakthroughs could drastically alter the industry's direction:

- **A Major War Between Great Powers**: If a conflict breaks out among leading countries, the arms industry would surge, fueling an arms race. But the devastation could be so great that the world calls for renewed disarmament afterward.
- **New Peace Movements**: A massive global peace movement might gain traction, leading to large-scale treaties banning entire classes of weapons. This happened in part with landmines and cluster munitions, though not all countries joined.
- **A Revolutionary Defense Breakthrough**: An invention that neutralizes or intercepts most offensive weapons (like a perfect missile shield) could upend warfare. Nations might try to replicate or surpass it, or it might push them toward negotiations if offense becomes futile.
- **Societal Shifts**: Younger generations, shaped by global connectivity and concerns like climate change, may reject large defense budgets and demand that resources go elsewhere. This cultural shift could put the arms trade under increasing pressure.

Such turning points remind us that the future is rarely linear. One major spark, whether peaceful or violent, can reorder how nations see the need for weapons and how the industry grows or shrinks.

---

## 12. Conclusion of Chapter 20

The weapon industry's future is tied to human choices about security, power, and cooperation. Several paths lie ahead:

- A new arms race fueled by technology like AI, quantum sensors, and space-based systems.
- Rapidly expanding private military firms, creating a world where corporate security can rival national armies.
- Greater reliance on non-lethal and defensive tools, perhaps guided by rising ethical demands.
- Increased militarization of space if orbital assets become prime strategic ground.
- Stricter treaties, disarmament, and transparency if global consensus decides advanced arms are too risky.
- Persistent small arms proliferation balanced by improved regulation and better law enforcement.
- Climate-driven conflicts that alter the nature of warfare and shift weapon designs.
- Public and investor pressure that reshapes corporate behavior and demands accountability.

Each scenario suggests different outcomes for the scale and nature of armed conflict. Though technology can push the industry in many directions, people's moral standards, governance, and collective problem-solving can steer it toward a safer world or a more dangerous one.

If humankind embraces collaboration and invests in peaceful solutions, the demand for lethal arms might gradually fade, or at least limit itself to strict defense roles. But if mistrust and competition win out, we could see an ever-rising tide of advanced and deadly tools. In the end, the future of the weapon industry reflects the broader values of society, reminding us that these technologies do not exist apart from human will—they are formed, sold, and deployed by people and for people, whether for good or ill.

# Help Us Share Your Thoughts!

**Dear reader,**

Thank you for spending your time with this book. We hope it brought you enjoyment and a few new ideas to think about. If there was anything that didn't work for you, or if you have suggestions on how we can improve, please let us know at **kontakt@skriuwer.com**. Your feedback means a lot to us and helps us make our books even better.

If you enjoyed this book, we would be very grateful if you left a review on the site where you purchased it. Your review not only helps other readers find our books, but also encourages us to keep creating more stories and materials that you'll love.

By choosing Skriuwer, you're also supporting **Frisian**—a minority language mainly spoken in the northern Netherlands. Although **Frisian** has a rich history, the number of speakers is shrinking, and it's at risk of dying out. Your purchase helps fund resources to preserve and promote this language, such as educational programs and learning tools. If you'd like to learn more about Frisian or even start learning it yourself, please visit **www.learnfrisian.com**.

Thank you for being part of our community. We look forward to sharing more books with you in the future.

**Warm regards,**
The Skriuwer Team

Printed in Dunstable, United Kingdom